Palgrave Studies in Literature, Science and Medicine

Series Editors
Sharon Ruston
Department of English and Creative Writing
Lancaster University
Lancaster, UK

Alice Jenkins
School of Critical Studies
University of Glasgow
Glasgow, UK

Catherine Belling
Feinberg School of Medicine
Northwestern University
Chicago, IL, USA

Palgrave Studies in Literature, Science and Medicine is an exciting new series that focuses on one of the most vibrant and interdisciplinary areas in literary studies: the intersection of literature, science and medicine. Comprised of academic monographs, essay collections, and Palgrave Pivot books, the series will emphasize a historical approach to its subjects, in conjunction with a range of other theoretical approaches. The series will cover all aspects of this rich and varied field and is open to new and emerging topics as well as established ones.

More information about this series at
http://www.palgrave.com/gp/series/14613

Hannah C. Tweed • Diane G. Scott
Editors

Medical Paratexts from Medieval to Modern

Dissecting the Page

Editors
Hannah C. Tweed
University of York
York, UK

Diane G. Scott
University of Glasgow
Glasgow, UK

Palgrave Studies in Literature, Science and Medicine
ISBN 978-3-319-73425-5   ISBN 978-3-319-73426-2   (eBook)
https://doi.org/10.1007/978-3-319-73426-2

Library of Congress Control Number: 2018934684

Cover illustration: Image by kind permission of University of Glasgow Archives & Special Collections, from their copy of Andreas Vesalius' 'De humani corporis fabrica' (SP Coll Bi6-a.5). Photograph by Robert MacLean.

This Palgrave Macmillan imprint is published by the registered company Springer Nature Switzerland AG
The registered company address is: Gewerbestrasse 11, 6330 Cham, Switzerland

Foreword

The notion 'paratext', the starting-point for the chapters in this collection, was famously enunciated by Gérard Genette back in the 1980s. A published text was to be studied in relation to the materials in which it was set, e.g. title-pages, frontispieces, dedications, forewords, prefaces, illustrations, foot-notes/endnotes, etc. Moreover, the notion of paratext could be extended to include such mediating phenomena as script, font, layout and punctuation, i.e. 'everything that originates in the sometimes very significant typographi-cal choices that go into the making of a book' (Genette 1997: 7).

As is now widely acknowledged, the 'text' does not have some abstract Platonic existence distinct from the contexts—including the paratexts—in which it exists at a given point in time. Classic textual criticism, as outlined by Karl Lachmann (1793–1851), presupposed the existence of such an 'ideal' or 'true' text, extracted 'from the manuscript evidence' (Reynolds and Wilson 2013: 223) in an attempt to 'restore the texts as closely as pos-sible to the form which they originally had' (Reynolds and Wilson 2013: 208). However, more recent scholarship has increasingly come to terms with the fact that texts are negotiable and negotiated through time, reworked through the copying process and situated in their own complex socio-cultural settings, however author-focused the enterprise of copying (by scribe or printer) or editing might claim to be (see further, Cerquiglini 1989). To put matters at their simplest: a modern edition of, say, Juvenal's *Satires*, however scholarly in presentation, simply cannot represent the 'form which [the texts] originally had', even if presented in *scriptio conti-nua* on a papyrus scroll. Readers or audiences in antiquity were very different from modern readers in how they engaged with the written word, and that different approach to textual engagement profoundly affected the textual form. Modern editors are increasingly conscious of such issues, even if there is ongoing dispute as to how to address them (see for medieval British studies, but with a wider resonance, Gillespie and Hudson 2013, and references there cited).

It is no surprise, therefore, that recent scholarship in the capacious field of textual cultures has emphasised what Vincent Gillespie has called 'total codicology': a 'highly detailed and finessed form of material and cultural history' that sees the 'text' in dynamic articulation with a broad range of other phenomena (Gillespie n.d.). In such research, the division between text and paratext and indeed wider cultural context is problematised: author, decorator/illuminator/illustrator, scribe/compositor, editor, reader/audience/annotator are all actors in the (ongoing) evolution of the work in question.

This articulation between textual form and textual function is at the heart of the chapters in the current collection: an exciting, forward-looking and necessarily wide-ranging contribution to interdisciplinary textual history, with special reference to medical humanities, that derived ultimately from a lively symposium devised and developed by two extremely impressive early-career researchers. It is a real pleasure to commend these chapters, and an honour to have been asked to contribute these opening remarks (themselves, of course, a paratext).

University of Glasgow   Jeremy J. Smith
Glasgow, UK
July 2017

BIBLIOGRAPHY

Cerquiglini, B. 1989. *Eloge de la variante: histoire critique de la philologie*. Paris: Seuil.

Genette, G. 1997. *Paratexts: Thresholds of Interpretation*. Trans. J. Lewin. Cambridge: Cambridge University Press.

Gillespie, V. n.d. http://www.lmh.ox.ac.uk/our-academics/fellows/prof-vincent-gillespie, consulted 11/7/2017.

Gillespie, V., and A. Hudson. 2013. Introduction. In *Probable Truth: Editing Medieval Texts from Britain in the Twenty-First Century*, ed. V. Gillespie and A. Hudson. Turnhout: Brepols.

Reynolds, L.D., and N.G. Wilson. 2013. *Scribes and Scholars: A Guide to the Transmission of Greek and Latin Literature*, 4th ed. Oxford: Oxford University Press.

Acknowledgements

This edited collection stems from a Wellcome Trust-funded conference of the same name, held at the University of Glasgow on 11 September 2015. First and foremost, sincere thanks are due to the collection's contributors for their critical engagement with the project—both those who participated in the original conference, and those who became involved later in the process. We would also like to thank Dr. Johanna Green (University of Glasgow) and Dr. Francesca Mackay (National Trust Scotland) for their enthusiasm, patience, and expertise in planning the conference, and to Professor Jeremy J. Smith for his ongoing support for our research into all things paratextual (and his willingness to provide the Foreword to the collection). Final thanks are due to Robert MacLean, Assistant Librarian at the University of Glasgow, and to the wider Archives and Special Collections team at the University of Glasgow Library, for providing access to their collections throughout the process, and tireless assistance with our queries and clarifications. Thank you!

This work was supported by the Wellcome Trust [WT109050AIA].

CONTENTS

NOTES ON CONTRIBUTORS

Natalie Calder received her AHRC-funded PhD from the School of Arts, English and Languages at the Queen's University of Belfast in 2017, where she is currently a Teaching Assistant. She has published on the influences of Lancastrian kingship on the religious poems of MS Digby 102, and historiographical treatments of medieval unbelief in Heresy Studies. Her doctoral research focused on modalities of belief and unbelief in late medieval lay devotional cultures, exploring the complex ways in which the laity of fifteenth-century England interrogated and engaged critically with Christianity. Her current research interests lie in medieval unbelief, historiography, spiritual despair and historical mental illness.

Elspeth Jajdelska is a Senior Lecturer in English at the University of Strathclyde. She is the author of two monographs on the changing relationship between speech and print in the seventeenth and eighteenth centuries. *Silent Reading and the Birth of the Narrator* (2007) explores the possibility that rising book ownership and changing attitudes to education created a critical mass of fluid, silent readers, which in turn led to changes in prose style and a reconceptualisation of the reader as a hearer, rather than speaker, of printed or written texts. *Speech, Print and Decorum in Britain, 1600–1750* (2016) identifies changing conceptualisations of written and printed documents across the period, from scripts and representations of speech to utterances in their own right, independent of their authors' bodies, and considers the implications of these changes for the relationships between rank, learning and freedom of speech.

Robert MacLean is an Assistant Librarian in the Archives & Special Collections of the University of Glasgow Library. He is primarily responsible for encouraging and facilitating use of the collections for learning and teaching and he has been involved in rare book cataloguing for more than fifteen years.

Laura Mainwaring obtained an MSc in the History of Science, Technology and Medicine from Imperial College in 2011. She is now studying for her PhD in the History of Medicine at the University of Leicester as part of a collaborative award with the Society of Apothecaries and the AHRC, focusing on the advertising, branding and packaging of healthcare products between 1870 and 1920.

Roberta Mullini is Professor of English Literature (now retired) at the University of Urbino Carlo Bo (Italy), and has published widely on English medieval and Shakespearean drama. She is also interested in theoretical issues connected to theatrical reception and to Shakespeare on screen. She has written volumes on First World War poetry (1977), on Shakespeare's fools (1983 and 1997), on late medieval plays (1992), on John Heywood (1997), on David Lodge's novels (2001), and on the material culture of the theatre (2003, with Romana Zacchi). She is now working on the aside in Shakespeare's plays. She has also directed students' performances of English interludes. Her book *Healing Words. The Printed Handbills of Early Modern London Quacks*, dealing mainly with linguistic aspects of medical practitioners' activities as documented in their advertisements, was published in 2015. She is editor-in-chief of *Lingua &*, a journal devoted to modern languages and cultures.

Harry Newman is a Lecturer in Shakespeare and Early Modern Literature at Royal Holloway, University of London. He publishes in the fields of early modern drama, book history, material culture, and the history of medicine. His first monograph, *Impressive Shakespeare: Identity, Authority and the Imprint in Shakespearean Drama*, will be published by Routledge in 2018. Other recent and forthcoming publications include "'[T]he Stamp of Martius'": Commoditized Character and the Technology of Theatrical Impression in *Coriolanus*" (*Renaissance Drama*, Spring 2017), 'Paratexts and Canonical Thresholds' (*Shakespeare*, 2017/2018), and a special issue on 'Metatheatre and Early Modern Drama', edited with Sarah Dustagheer (*Shakespeare Bulletin*, Spring 2018). He runs The Paper Stage, an early modern play-reading series for students, researchers and members of the public.

Louise Powell is a PhD candidate in English at Sheffield Hallam University. Her doctoral research is fully funded by NECAH (North of England Consortium for Arts and Humanities Research). Her thesis explores how seventeenth-century medical and dramatic representations of twins reveal concerns about the masculinities of surgeons, playwrights and performers. Her research has been featured as part of the Wellcome Trust's 'Spotlight Researcher' series, and published in *Early Modern Literary Studies* and *Parergon*. She is the Yorkshire and the Humber representative for the Arts Health Early Career Research Network, and reviews Shakespeare's comedies for *The Year's Work in English Studies*.

Diane G. Scott received her PhD, on late medieval book history, from the University of Glasgow. She is the Research Associate for the AHRC Digital Transformations Theme and teaches in the Department of English Language and Linguistics at the University of Glasgow. Her research focuses on fifteenth- and sixteenth-century literacy and literary culture.

Deborah Thorpe is a Trinity Long Room Hub Marie Sklodowka-Curie Co-fund Fellow at Trinity College Dublin and a Visiting Fellow at the University of York. Her current research involves palaeographical study of the work of ageing and elderly scribes in medieval and early modern manuscript books and documents. The overarching aim of this project is to better understand the handwriting changes associated with normal physiological ageing, as well as the stylistic developments that occurred as fashions changed, and scribes were influenced by patrons and other scribes around them. Her work for the chapter in this volume was part-funded by the Wellcome Trust [ref: 105624] through the Centre for Chronic Diseases and Disorders (C2D2) at the University of York.

Hannah Tweed received her PhD from the University of Glasgow, on representations of autism in contemporary literature and film. She is a Postdoctoral Research Associate at the University of York, on the 'Cultures of Care' project. Her research focuses on twentieth- and twenty-first-century literature, with specialisms in disability studies and the medical humanities.

LIST OF FIGURES

CHAPTER 1

Authority, Authenticity and Reputation: An Introduction to Medical Paratexts

Hannah C. Tweed and Diane G. Scott

CASE STUDY: MARY TOFT

In September 1726, John Howard, a surgeon from Guildford, witnessed Mary Toft deliver nine rabbits. By October, Toft was being visited by members of the public and medical professionals alike as a medical curiosity. Her case was discussed and dissected in newspapers, printed ballads and satirical sketches, and Toft's predicament was considered by many to be an example of the maternal imagination—her rabbit births the result of eating and craving rabbit meat while pregnant (Lieske 2007: xiii). In November, following three months of similar sporadic 'births' and increasing public interest, Nathanael St. André, surgeon and anatomist to the Prince of Wales, and Cyriacus Ahlers, surgeon to George I's German household, visited Toft in Guildford, where she was again observed delivering parts of a rabbit (Todd 1995: 26–27). Suspicious of fraud, Ahlers'

H. C. Tweed (✉)
University of York, York, UK

D. G. Scott
University of Glasgow, Glasgow, UK

© The Author(s) 2018
H. C. Tweed, D. G. Scott (eds.), *Medical Paratexts from Medieval to Modern*, Palgrave Studies in Literature, Science and Medicine,
https://doi.org/10.1007/978-3-319-73426-2_1

1

reports led fashionable physician Richard Manningham to examine Toft, delivering an object subsequently identified as a hog bladder. Toft was finally moved to London and observed by the trio of surgeons and physicians, along with new specialist 'men-midwives', such as James Douglas, while being pressured to admit fraudulent delivery. Eventually, following evidence that Toft's husband and a porter were supplying her with dead rabbits (and Manningham's threat to perform a forced caesarean section), Toft finally confessed to the fraud.

Douglas' published response, *An Advertisement Occasion'd by Some Passages in Sir R. Manningham's Diary, Latedly Publish'd, by J. Douglas, M.D.*,[1] presented a copy-edit of Manningham's publication, querying the accuracy of certain sections and reiterating his own claims to have never been convinced by Toft, as well as detailing her symptoms and habits. Yet while the content of these two texts might indicate an intended audience of medical professionals, Toft's melodramatically entitled *Much ado about Nothing: Or, The Rabbit-Woman's CONFESSION*, and popular ballads such as *The DISCOVERY: or, The Squire turn'd Ferret. An Excellent New BALLAD* demonstrate how the commentaries were titled with a view to public consumption as much as (if not more than) medical professionals. Mary Fissell's work on the politics of reproduction in early modern England draws connections between the availability of monstrous birth narratives in cheap printed pamphlets and a cultural shift in the way the womb was viewed across society, both medical and lay. In a post-Reformation society, women were no longer encouraged to identify with the Virgin Mary during pregnancy, and the womb, previously a site of miraculous life-giving, was increasingly viewed as 'the source of many women's maladies'. Cheap print allowed for the proliferation of sensational stories, and women's bodies, pregnancy and motherhood were popular choices (Fissell 2006: 53).

The Toft affair certainly highlights developments in print technology and the commercial possibilities it afforded eighteenth-century printers and publishers; more importantly, however, it also demonstrates how the eighteenth-century medical marketplace was able to exploit the *promotional* potential of print. The physicians involved in the case were keen to document their work and observations, and some made their way quickly into print. Both *The Wonder of Wonders*,[2] which contains an excerpt of correspondence by surgeon John Howard, and *A Short Narrative*,[3] written by Nathaniel St André, were published shortly before Toft's confession (Lieske 2007: 21). In addition to providing details of the surgeons' actions in the case, both texts attempt to quell doubts over the authenticity of the births and 'counter an anticipated assault on [the physicians'] public and professional image' (Lieske 2007: xii).

That so many well-respected medical professionals were fooled or even implicated in the fraud triggered further waves of published material relating to the affair. In the months following Toft's confession, the physicians, surgeons and midwives involved in the case published a number of pamphlets and accounts defending their own professional reputations (and often laying blame on other parties). These publications were undeniably sensational in nature, as well as deeply intertextual—requiring and expecting from their readers a substantial knowledge of the case and the associated medical details. Manningham's published diary extracts, for example, present the case in explicit medical detail, while also cementing his claims to have never been convinced by the births.[4]

Mary Toft's case engaged with a range of eighteenth-century concerns by and about the medical profession: the role of sensationalism in professional practice; public questions about medical authority and verisimilitude; the accountability of medical men to the public (of all classes). As Pamela Lieske states, however, the Toft case 'did not occur in a cultural vacuum, but [is] part of a larger historical tradition of strange and monstrous births, whose meaning contemporary scholars are still attempting to decipher' (Lieske 2007: viii). Situating the Toft affair within a history of 'monstrous births' which stretches back to the medieval and classical periods, Lieske argues that such cases should not be dismissed as mere fantasy, but that we should consider how the medical profession made 'concerted efforts [. . .] to make sense of what they observed and experienced' (Lieske 2007: viii). Mary Toft may have been at the centre of the affair, but her female body quickly became a physical and rhetorical battleground between the male medical professionals defending their reputations and livelihoods.

The Mary Toft case has long been noted as a bizarre but intriguing example of late eighteenth-century medical and gynaecological practices[5] and, from a book history perspective, of the role of the popular press during this time. This chapter, however, aims to draw attention specifically to the paratextual features of the various publications. Paratext constitutes the linguistic and visual features which are contained within and surround the main text(s); meaning and information are communicated by and encoded in these features.[6] More specifically, the various Toft publications are examples of the 'medical paratext' and highlight the breadth and complexity of material which this term may encompass. In addition to demonstrating wider concerns and preoccupations within the eighteenth-century medical profession, the Toft affair texts and their paratexts also reflect contemporary public opinion about physicians and their trade. A focused investigation of the paratext not only highlights the complexities of the case itself, it provides insight into the structures and knowledge base of the medical profession in the late eighteenth century. Furthermore, the various elements of paratextual detail and the distinctive inter(para)textuality expose the tensions between public and professional, medical and lay, and privacy and sensationalism.

Paratext is always a site of mediation between the reader, printer/publisher, editor, and author; it is often also a site of tension.[7] This tension is clearly evidenced in the Toft publications, as medical professionals used the printed text and its paratextual framing devices to promote their expertise, defend their reputations and demonstrate accountability to the reading public. Furthermore, many of the texts were produced in direct response to each other, and to developments—allegations and revelations—over the course of the affair. As such, the texts, and specifically the paratexts, are dialogic. Titles such as *An Advertisement Occasion'd by Some Passages in Sir R. Manningham's Diary, Latedly Publish'd, by J. Douglas, M.D.*, a tract in which Douglas makes 'remarks' upon Manningham's account, are explicit in their inter(para)textuality. Acknowledging the paratext, then, not only demonstrates how medical professionals responded to the case, but how they engaged with each other, and to the wider reading public.

The voices of the male physicians take centre stage across the publications but the voice of Mary Toft is notable in its absence; when we do 'hear' her voice, it is heavily mediated. The following text comprises the opening pages of Toft's printed confession, *Much ado About Nothing*, which features an explanatory preface from the publisher:

The Publisher to the Reader

The poor woman of *Godalming* being now the topic of every conversation, and it being put to the general vote, whether rabbits shall be admitted to our tables, ay or no; it has been thought fit to trace the whole affair from its first original; and to hear what the poor woman has to say for herself, at a time when all mouths are open against her: in order to which, the publisher hereof has taken indefatigable pains to bring the whole mystery to light, by purging the woman in a proper manner, and at proper times, without the low artifice of wheedling, or the high hand of threatening; but by touching *her in the tenderest part*, vis. her conscience; and extracting the very quintessence of the whole affect in such a manner, and method, as will set all mankind to rights in their various mistaken notions of this unhappy woman.

It is therefore to be hoped, they will suspend their judgements, till they have heard what she has to say for herself; and that upon a mature recollection and debate of the whole, they will set the saddle on the right horse, by letting their resentments fall on *the true imposters, or quacks,* and not on a poor innocent woman, whose misfortunes they have made the cat's paw of their roguery.

Postscript.

It was thought fit to print this confession in *puris naturalibus,* (i.e.) in her own stile and spelling, without any amendment or adulteration, which would but spoil its natural simplicity, and render it less genuine and credible.[8]

On the surface, the Preface extends a great deal of sympathy to the 'poor innocent woman' and laments the infamy which her predicament has brought upon her. The reader is urged to refrain from judgement until they have 'heard' her own explanation of events which will, the Preface assures us, reveal the 'quacks' to be at fault. However, an unmistakable mocking tone undermines the attempt at a sympathetic portrayal of Mary Toft. The Preface includes a thinly veiled reference to Toft's vagina ('tenderest part'), and perhaps a pun on the 'indefatigable pains' of labour. The decision to print the confession in Toft's 'own stile', apparently without editorial input, also requires interrogation. The claim that Toft's words are presented without 'amendment' is contradicted by the very process of mediation which takes place in the Preface. The publisher has deliberately framed Toft's 'confession' and such framing devices are always used to control, or at least direct, the reader's interpretation and reading process.

Attempts to control the narrative of events and direct reader engagement through the use of paratext is a prominent feature of much of the contemporary Toft material. As we have noted, however, the attempts to seek the truth were as much about personal reputation as they were about scientific authority and professional accountability. Furthermore, the surgeons and physicians were attempting to exercise a level of control within a medical marketplace which offered readers, both medical and lay, increasing access to cheaply printed books and pamphlets. This level of access posed a particular challenge to medical authority during a time when the concept of medical authority was continuing to shift towards professionalisation.

Dissecting the Page

Bonnie Mak describes the page as 'an interface, standing at the centre of a complicated dynamic of interaction and reception'; the present collection examines how this interface functions within and around medical texts (Mak 2011: 21) By uniting pragmatics-based analysis with the history of medicine, the collection and its contributors seek to (re)construct the interactions between the medical text and its producers and users.

While many of the chapters engage in detailed synchronic analysis of a specific text or time period, the collection as a whole is diachronic, spanning the fourteenth to the twenty-first centuries. The combined synchronic and diachronic perspectives offered herein provide both qualitative analysis of texts and their readership, and an opportunity to chart the development of medical paratexts from script to print. This collection offers insight into the range of methodological and theoretical approaches to medical paratexts currently employed by researchers working across a variety of disciplines and time periods.

The following chapters open up discussions of medical paratexts in a number of ways, beginning with how we might define (or not) the boundaries of the paratextual domains as they relate to the history of medicine, and to the production and reception of medical texts. Several of the chapters focus primarily on the domain of *peritext*, analysing features such as illustrations and annotations found within the confines of the material text. These chapters are concerned with evidence of the circumstances of production, authorial and/or editorial intentions, and reader engagement and response. Other contributors draw our attention to the wider domain of *epitext*, moving beyond the text proper and considering the various social, political and cultural factors which shape the production of medical(ised) texts and inform the producers and consumers of those texts.

The study of paratexts allows the marginal to take centre stage; the liminal spaces and 'thresholds' described by Gérard Genette become important sites for the generation of meaning (Genette 1997). This collection demonstrates and discusses a variety of potential sites and offers tools for the excavation of pragmatic meaning within medical texts. The reconsideration of the marginal is particularly relevant within medical humanities discourse, in which issues of authority, autonomy, and ethics are at the forefront of the field. Not only does this collection engage with features traditionally considered marginal and/or ephemeral, the very act of uncovering layers of meaning encoded in the linguistic and visual features of the text allows us to highlight marginalised uses and users of those texts. A strength of this collection, therefore, is its focus on the lay users and consumers of medical texts, in addition to the medical practitioners and producers.

The chapters in the collection span from the medieval to the modern, with a particular focus on the seventeenth century. The sixteenth and seventeenth centuries saw significant progress in the understanding of human anatomy (Andreas Vesalius, *De Humani Corporis Fabrica* (1543); Ambroise Paré, *Les Oeuvres* (1575); William Harvey, *Du Motu Cordis* (1628)). Medical historian Roy Porter states that by 1700 'advances in gross anatomy – and, after William Harvey, in physiology also – had created the dream of a scientific understanding of the body's structures and functions, drawing on and matching those of the new and highly prestigious mechanics and mathematics' (Porter, 2006b: 142). Similarly, 'surgery was showing signs of becoming more methodical' (Porter 2006a: 179), with 'specialist surgeons pioneered new techniques […] Thanks to improvements in surgery and changes in obstetrical practices, surgery rose in professional standing' (Porter 2006a: 192–3). In Elspeth Jajdelska and Roberta Mullini's chapters, handbills and advertising materials for surgeons and physicians are shown to engage explicitly with the changing public status of medical practitioners via the *peritext* and *epitext*, as surgeons and physicians attempted to reassure the public of their professional status (as in the Toft case). These medical paratexts represent a touchstone for popular culture as it interacts with scientific and medical developments—a pattern that continues into the modern period. Thus, two of the final chapters in the collection focus on twentieth-century texts that destabilise gendered ideas of medical authority and role of the medical 'expert' ("'Nonsense Rides Piggyback on Sensible Things': The Past, Present, and Future of Graphology" by Thorpe, and "Archives, Paratext and Life Writing in the First World War" by Tweed).

This collection proposes to establish the term 'medical paratexts' as a useful addition to medical humanities, book history, and literary studies research. Existing scholarship, such as Elizabeth Lane Furdell's *Textual Healing* (2005), or Irma Taavitsainen and Päivi Pahta's *Early Modern English Medical Texts* (2010), has focused on the connections between the history of medicine, pragmatics-based analysis, and texts produced by medical practitioners. This collection is also text-focused, and introduces significant archival material. However, our contributors have expanded the understanding of medical paratexts to include wider readerships and consumption of medical texts, and particularly the interactions between readership and text production and reception.

In choosing to focus on reader response and book history, the chapters in this collection display provocative and significant overlaps between lay and expert opinions in medicalised texts. Key themes emerge from this focus—in particular, the pervasive use of paratexts to question and destabilise authority (of medical professionals, of social structures, of gender bias). In a field where concern over the possession and use of bodies is a prevalent theme, paratexts reflect wider anxieties about medicine, science, and culture.

Summary of Chapters

Dissecting the Page is structured into two thematic Parts, underpinned by a shared examination of ideas of medical and lay readership and a history of reader response. Part I, 'Production, Reception, and Use' focuses on the production, reception, and use of medical texts—particularly texts that claimed to be primarily for medical professionals. Part II, 'Authority, Access, and Dissemination' analyses the role and significance of authority, access, and dissemination in discussions of health, medicine and illness, for both lay and medical readerships.

Part I: Production, Reception, and Use

Harry Newman's chapter, ""[P]rophane fidlers': Medical Paratexts and Indecent Readers in Early Modern England", focuses on late sixteenth- and early seventeenth-century texts. Newman analyses so-called 'Prophane Fidlers' in his examination of medical paratexts and indecent readers in early modern England. This period saw a proliferation of printed vernacular texts that discussed and illustrated female reproductive organs. Newman considers how prefatory writers—not just authors but also translators and publishers—justified their publications to 'legitimate readers' (modest women and medical professionals) and admonished the intrusiveness of 'illegitimate readers' (laymen). Newman illustrates how these writers employed rhetorical strategies that both established the publications' legitimacy and fetishised the books as marketable erotic objects to lay readers.

Louise Powell's chapter "Touching Twins in the Texts and Medical Prefaces of Seventeenth-Century Midwifery Books" examines two seventeenth-century midwifery books, with a specific focus on their representations of twins. Powell argues that the written texts and illustrative paratexts present complex and contradictory representations of twinship in the early modern period. Her exemplar texts and paratexts present very different ideas about the ways in which twins were understood by medical practitioners in the early modern period.

Powell's essay, with its focus on medical illustrations, overlaps usefully with Roberta Mullini's chapter, "Graphic Surgical Practice in the Handbills of Seventeenth-Century London Irregulars". Mullini outlines how in the second half of the seventeenth century the London medical marketplace saw an increase in the publication of advertisements aimed at selling products for healthcare, and highlighting practitioners' surgical skills. This chapter examines the structure of the handbills of different medical irregular practitioners. Mullini then analyses what kind of dialectics—if any—are detectable between verbal and visual messages, and the early modern multimodal power of these handbills. Also focusing on the relationship between illustrations, advertisements, and readers is Laura Mainwaring's chapter "Profit and Paratexts; the Economics of Pharmaceutical Packaging in the Long Nineteenth Century" on advertising and packaging in the long nineteenth century. Concentrating on pharmaceutical packaging between 1860 and 1920, Mainwaring's chapter broadens the understanding of paratext to include the materials and marks found on packaging. She argues against traditional explanatory models of the nineteenth-century marketplace, which have dismissed consumers as gullible fools duped by patent medicine vendors. Through close readings of the packaging, Mainwaring demonstrates that wholesale and manufacturing firms considered their consumers more adept at decision-making than the traditional narrative has allowed.

Part II: Authority, Access, and Dissemination

Natalie Calder's chapter, "Remedies for Despair: Considering Mental Health in Late Medieval England", focuses on the instructions for meditative exercises given in William Bonde's *Treatise for them that ben tymorous and fearefull in conscience* (1527). Calder suggests that Bonde's *Treatise* links devotional practices and theological instruction to mental health care, acting as an accompaniment to general manuals that dealt with the care of the soul and spiritual life. This focus on potentially liminal paratexts—the wider reader environments established by Bonde's advice—and the overlap between religion and health link to Elspeth's Jajdelska's chapter, "The Medical Paratext as a Voice in the Patient's Chamber: Speech and Print in *Physick* for the Poor (1657)". Jajdelska asks what can rhetorical approaches in paratexts tell the reader about the different social relationships possible between men of medicine and their patients of varying ranks, focusing particularly on the role of the clergy. She proposes that the selected paratexts can be seen as representations of speech, and can be reinterpreted as analogues for spoken encounters.

Moving from public speech acts to private diaries, Hannah Tweed's chapter "Archives, Paratext and Life Writing in the First World War" focuses on critically neglected female nurse-writers from the First World War, and their implicit audiences. Concentrating on diarists Alice Lighthall and Claire Gass, Tweed suggests that the *epitext* and *peritext* were central to the diarists' writing process and are subsequently key to reader engagement with these narratives. Furthermore, Tweed argues that these nurse-writers were self-conscious archivists, evident in the form, function, and interaction of the paratextual detail within their diaries.

The most contemporary contribution to the collection takes an exploratory approach to what should be considered as paratext. In "'Nonsense Rides Piggyback on Sensible Things': The Past, Present, and Future of Graphology", Deborah Thorpe considers the state of graphology in the late twentieth and early twenty-first centuries. She focuses in particular on how the study of handwriting has been appraised by medical practitioners, scientists, and lay readers. Thorpe examines the potentially damaging impact of graphology, highlighting the anxiety that judgements about handwriting can cause to individuals with movement disorders. She asks what can we learn about personality – or states of being – from handwriting, and how handwriting features intersect with symptoms of disease and disorder.

The final chapter of the collection focuses on archival holdings and medical paratexts. Robert MacLean ends the collection with an introduction to medical marginalia in the early printed books held by the University of Glasgow's Special Collections. This chapter provides a collections-based approach to a wide range of examples of medieval and early modern medical marginalia, and a thoughtful response to some of the issues faced by archives and archivists in the digital age.

CONCLUSION

Mary Toft's case, and the publications surrounding it, illustrate how concerns relating to the professionalisation of medicine and developments in scientific and medical knowledge during the eighteenth century were played out in the paratexts of these publications. The layers of meaning and tension are communicated by the interaction of the various textual and paratextual elements; the tension is in the interplay. Only by unpacking these various levels and features, and acknowledging the tensions and contradictions therein, can we gain an insight into the complexities of medical practice, attitudes, authority, and reader engagement—both medical and lay—and how they have developed over time.

This edited collection proposes that the term 'medical paratext' is an important and timely addition to discussions—social, cultural, and medical—surrounding medical or medicalised texts. The contributors to this collection analyse medical texts ranging from the medieval to the modern. The primary material includes advertising, case notes, autobiographical writing, and marginalia by patients and medical practitioners. The collection thus engages with the complex interactions between the content and paratextual information found within non-fictional medical texts, bringing to the fore shifting notions of medical authority and the (often problematic) role of the reader/patient.

NOTES

1. *An Advertisement Occasion'd by Some Passages in Sir R. Manningham's Diary, Latedly Publish'd, by J. Douglas, M.D.* (1727), printed by J. Roberts, London (ESTC T056026).
2. *The Wonder of Wonders: Or, A True and Perfect Narrative of a Woman near Guildford in Surrey, who was Delivered lately of Seventeen Rabbets* (1726), printed by J. Bagnell, Ipswich (ESTC T055627).
3. *A Short Narrative Of an Extraordinary Delivery of Rabbets* (1727), printed by John Clarke, London (ESTC T055617).
4. *An Exact Diary Of what was observ'd during a Close Attendance Upon Mary Toft, The pretended Rabbet-Breeder of Goldaming in Surrey* (1726), printed by Fletcher Gyles, London (ESTC T056206).
5. An entire volume of the *Eighteenth Century Midwifery* series is devoted to the Mary Toft case.

6. Following traditional paratextual categorisation and description, we might consider the linguistic and visual features contained within the text as the *peritext*, information which 'one can situate in relationship to that of the text itself [...] around the text, in the space of the same volume'—prefaces, explanatory notes, appended correspondence, and reader additions (Genette 1991. 263). The *ɵiptext* constitutes 'any paratextual element not materially appended to the text within the same volume' which might circulate in 'physical and social space' (Genette 1997: 344).

7. For an in-depth discussion of the politics of paratext, see *Renaissance Paratexts*, ed. by Helen Smith and Louise Wilson (Cambridge: Cambridge University Press, 2011).

8. *Much ado about nothing: or, a plain refutation of all that has been written or said concerning the rabbit-woman of Godalming* (1727), printed by A. Moore, London (ESTC T055626).

BIBLIOGRAPHY

Fissell, M.E. 2006. *Vernacular Bodies: The Politics of Reproduction in Early Modern England*. Oxford: Oxford University Press.

Genette, G. 1991. Introduction to the Paratext. *New Literary History* 22 (2): 261–272.

———. 1997. *Thresholds of Interpretation*. Cambridge: Cambridge University Press.

Lieske, P., ed. 2007. *Eighteenth Century Midwifery*. Vol. 2. London: Pickering and Chatto Publishers.

Mak, B. 2011. *How The Page Matters*. Toronto: Toronto University Press.

Porter, R., ed. 2006a. Hospitals and Surgery. In *The Cambridge History of Medicine*, 176–210. Cambridge: Cambridge University Press.

———., ed. 2006b. Medical Science. In *The Cambridge History of Medicine*, 136–175. Cambridge: Cambridge University Press.

Smith, Helen, and Louise Wilson, eds. 2011. *Renaissance Paratexts*. Cambridge: Cambridge University Press.

Todd, D. 1995. *Imagining Monsters: Miscreations of the Self in Eighteenth-Century England*. Chicago: University of Chicago Press.

PART I

Production, Reception, and Use

'[P]rophane fidlers': Medical Paratexts and Indecent Readers in Early Modern England

Harry Newman

Michael Camille's work on the 'interpenetration of corporeality and codicology' in the medieval period suggests the fetishization of books long pre-dates the early modern period (Camille 1997: 35). But the creators of paratexts in sixteenth- and seventeenth-century England were remarkable for their tendency to conflate printed books and sexed bodies. Authors, translators and stationers framed books in ways that made it difficult not to conceive of their handling and marking by readers as indecent. A preface by the printer-publisher John Day to his 1570 edition of *The Tragidie of Ferrex and Porrex*, by Thomas Norton and Thomas Sackville, denounces an earlier, supposedly unauthorised edition (1565) as corrupt by comparing its publisher to one who has 'enticed into his house a faire maide and done her villainie, and after all … bescratched her face, torne her apparel, berayed and disfigured her, and then thrust her out of dores dishonested' (A2r). Printed books were represented as not just deflowered or raped virgins, but also prostitutes, adulterous wives, and bastard children. In *The*

H. Newman (✉)
Department of English, Royal Holloway, University of London, Egham, UK

© The Author(s) 2018
H. C. Tweed, D. G. Scott (eds.), *Medical Paratexts from Medieval to Modern*, Palgrave Studies in Literature, Science and Medicine,
https://doi.org/10.1007/978-3-319-73426-2_2

15

Imprint of Gender, Wendy Wall traces 'a pervasive cultural phenomenon in which writers and publishers ushered printed texts into the public eye by naming that entrance as a titillating and transgressive act'. Such rhetorical strategies, she argues, were used to encode and negotiate readers' and writers' class-based anxieties about printed publication (Wall 1993: 172–73). More recently, Michael Saenger has argued that textual/sexual tropes served a wider prefatory advertising strategy of 'metaphoric induction', whereby publishers 'implicate a variety of perusers into potent imaginary relationships which continue through the point of purchase, with the promise of a differed satisfaction … through private reading' (Saenger 2006: 96–97).

Wall and Saenger are more interested in 'literary' than 'medical' paratexts (an opposition I hope to trouble), but many of the books being prefaced by images of sexually transgressive disclosure in early modern England contained information and advice about sex and reproduction. The late sixteenth and seventeenth centuries saw a proliferation of printed vernacular gynaecological and obstetrical texts that discussed and illustrated the female reproductive organs, often referred to as women's 'secrets' or 'privities'.[1] Exclusively male-authored until the late seventeenth century, these texts included midwifery manuals (also discussed in Louise Powell's chapter "Touching Twins in the Texts and Medical Paratexts of Seventeenth-Century Midwifery Books" in this volume), and sections within large anatomical works, such as Book 4 of Helkiah Crooke's *Mikrokosmographia* (1615). The publication of these books inflamed moral outrage, even within the medical establishment, and they were stigmatised as a kind of pornography, texts whose existence meant that the mysteries of the female body were up for general sale.[2] Those who wrote and published on what Crooke calls women's 'obscoene parts' (200) worked in a context where the connection between the printed book and the sexed female body was more than just figurative.[3] They did so, furthermore, in a culture that recognised reading as an embodied activity that could stimulate a wide range of physiological processes in men and women.[4]

While literary, social and medical historians have addressed the relationship between medical and erotic literature in early modern England, little has been written on the significance of book history to such discussions.[5] In particular, paratexts can tell us much about the connections between the rhetorical or 'literary' qualities of these books and their material life during publication and reading processes.[6] Using Book 4 of Crooke's *Mikrokosmographia* (dedicated to the reproductive organs) as a case study, this chapter combines archival work with analysis of rhetoric and illustrations to investigate how printed paratexts to these kinds of books (e.g. title-pages, dedications, prefaces) attempted to manipulate responses and construct 'implied' readers of two camps: the modest lay-women, midwives and male medical professionals to whom the texts were addressed ('legitimate' readers), and laymen whose readership was scorned ('illegitimate' readers). In analysing these paratexts' rhetoric and their 'implied' readers, I compare and contrast the identities and activities of 'actual' readers who produced their own paratexts in the form of inscriptions, annotations and illustrations.[7] I argue that the creators of printed paratexts employed rhetorical, visual and material strategies which, while explicitly establishing the publications' legitimacy, fetishised the books as marketable erotic objects, making them more attractive to consumers driven by prurient curiosity. And in my focus on 'actual' reader responses, I suggest the manuscript paratexts produced by an apparently diverse range of readers can be interpreted as both responding and contributing to the perceived erotic agency of these texts. More broadly, I hope to posit early modern medical texts as trans-historical objects whose on-going semiotic transactions with readers depend on the relationship so often negotiated by paratexts: that between the rhetorical and the material.

In what follows, I discuss the publication and readership of *Mikrokosmographia*, before analysing—in turn—prefacing authors' use of the rhetorical genre of *apologia* (addressed to 'legitimate' readers) and the figure *admonitio* (for 'illegitimate' readers). In doing so, I connect a range of paratexts—print and manuscript, visual and textual, peritextual and epi-textual—that offer insights into these medical texts' production, circulation and reception.[8]

MIKROKOSMOGRAPHIA: PUBLICATION, CENSORSHIP AND READERS' MARKS

Over a thousand folio pages in length, *Mikrokosmographia: A Description of the Body of Man*, written by the London physician Helkiah Crooke (1576–1648),[9] was the first comprehensive anatomical work in English. It was printed and published in 1615 by William Jaggard, who re-issued it in 1616 and 1618 (STC 6062, 6062.2, 6062.4; Russell 1963: 218–22).[10] A second edition—revised and expanded by Crooke—was printed in 1631 by Thomas and Richard Cotes for Michael Sparke (STC 6063; Russell

1963: 223). A third edition—printed by Richard Cotes for John Clarke after Crooke's death—was released in 1651 (Wing C7230; Russell 1963: 224). The volume draws heavily on Latin works by the French physicians Caspar Bauhin (1560–1624) and André du Laurens (1558–1609), and has a large number of elaborate woodcut illustrations, the majority of which originate from Andreas Vesalius' Latin anatomical work of 1543, *De humani corporis fabrica libri septem*, but were transmitted—mostly if not entirely—via other texts. A number of the illustrations, for instance, seem to have come from Bauhin's *Theatrum anatomicum* (1592, 1605), one of Crooke's acknowledged sources (Linster 2017: 55, 77).

Mikrokosmographia's illustrations of the female body were central to the controversy of its publication in 1615, as evidenced by the Annals of the Royal College of Physicians.[11] Primarily at issue were Book 4, on "the Naturall Parts belonging to generation, as well in Men as in Women" (Crooke 197), and Book 5, an obstetrical text "Wherein the Historie of the Infant is accurately described" (257). After parts of the volume were released in 1614, the College of Physicians, headed by its president William Paddy, shared objections voiced by the Bishop of London John King (responsible for licensing), and in particular condemned Book 4 and its 'indecent illustrations' (Annals, 11 November 1614). Five months later Paddy declared that if the volume was published with Book 4 unchanged, 'he would burn it wherever he might find it' (Annals, 3 April 1615). Ultimately, however, efforts to suppress or censor the publication in 1614–15 were apparently unsuccessful: an interpolation in the annals notes 'the whole book' was published, and the title-page—in a variant issue of the 1615 edition (Russell 1963: 219)—states it has been published '*according to the first integrity, as it was originally written by the Author*'.[12]

Crooke addressed the controversy surrounding the publication in several of the first edition's prefatory texts. All thirteen books of *Mikrokosmographia* have individual prefaces by Crooke, and the work as a whole is introduced by the author's Latin dedication to King James I (¶3r–4v), his long English epistle to the Company of Barber-Surgeons (¶1r–2v), and—on an added bifolum—a dedicatory poem by Ambrose Fisher in Greek, Latin and English (¶4r). Crooke's letter to the surgeons addresses five 'obiections' made against him: 1.) for writing in English instead of Latin,[13] 2.) for not being original, 3.) for not writing well, 4.) for using illustrations as 'obscoene as *Aretines*' (i.e. Pietro Aretino, an Italian writer famous for his *Sonetti Lussuriosi* (1527), published with pornographic drawings by Giulio Romano)[14] and 5.) for including discussions of the reproductive organs at all (¶2v). Crooke responds most comprehensively to the last objection, especially in his preface to Book 4, which I will discuss in detail later. But it is significant that Crooke's refusal to remove any 'indecent illustrations', which he declared to be 'no other then those of *Vessalius, Plantinus, Platerus, Laurentius, Valuerdus, Bauhinus,* and the rest' (¶2v), would have satisfied Jaggard's desire to protect his large investment in woodcuts,[15] illustrations whose association with Italian pornography might have improved *Mikrokosmographia*'s commercial success.

The volume's illustrations were even published separately by Jaggard the following year in an octavo epitome, *Somatographia Anthropine*, prefaced by the Scottish surgeon Alexander Read (1616; reissued by Michael Sparke in 1634; STC 20782, 20783; Russell 1963: 681, 682).[16] Read's address "To the Courteous Reader" advertised its wider appeal and accessibility, observing not only that the 'portable' epitome could be 'carried without trouble, to the places appointed for dissection', but also that it would prove 'profitable and delightfull to such as are not able to buy or haue no time to peruse the other [i.e. *Mikrokosmographia*]', having 'proceede[d] from a mind desirous to give satisfaction to all' (A4v). That mind seems to have belonged to Jaggard or Read rather than Crooke, who was apparently not involved in the epitome's compilation or publication. Nonetheless, *Somatographia*'s small size, relative cheapness and privileging of image over text made material from *Mikrokosmographia*—including its 'indecent illustrations'—accessible to the book-buying public in a whole new way. '[T]his small volume', declares Read in a statement that probably horrified William Paddy, 'impresseth the Figures firmely in the mind' (A4v).

At least one of the figures, however, was now different. A remarkable image variant, discovered by Jillian Linster, indicates an act of censorship during or prior to the production of the epitome in 1616.[17] In *Somatographia* and its re-issue, a Vesalian-imitation illustration[18] of an anatomised female torso from *Mikrokosmographia*'s first edition (see Fig. 2.1) has been altered: the vulva has been erased, apparently by taking a knife to the detail in the woodcut but leaving behind the characters that label it ('y y'). While the uncensored woodcut of *Mikrokosmographia*'s first edition appeared again in the re-issues of 1616 and 1618, the censored version was re-used for the 1631 and 1651 editions (see Fig. 2.2).[19] It is unclear why the image was changed in 1615/1616 and yet remained unaltered in the first edition re-issues, and we cannot be sure whether

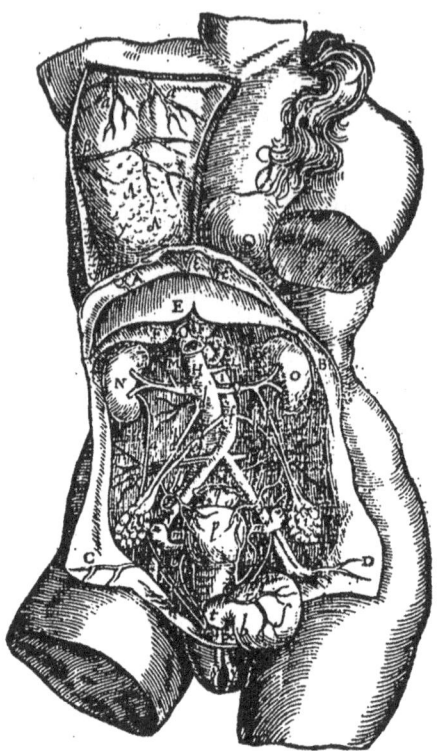

Fig. 2.1 Uncensored version of anatomised female torso in the 1615 edition of *Mikrokosmographia* (219) (RB 53894, The Huntington Library, San Marino, California. Image produced by ProQuest as part of *Early English Books Online*. <www.proquest.com> Image published with permission of ProQuest. Further reproduction is prohibited without permission.)

the alteration happened with the knowledge or approval of Crooke or even Jaggard.[20] What is striking about the censored image, however, is the potential that the negative space produced by the erasure has to generate greater interest in a part of the female body that becomes all the provocative for its absence.[21] It is, in effect, a fetishised absent-presence, and all the more so because the image's accompanying text (in all editions) tells the reader what is supposedly missing: 'yy Certaine skinnie Caruncles of the priuities in the middest of which is the slit, and on both sides appeare little hillocks' (Crooke 1631: 218). *Somatographia* refers

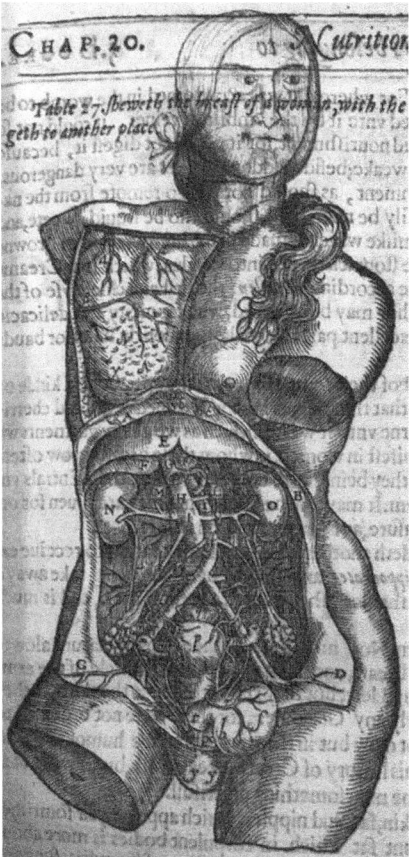

Fig. 2.2   Censored version of anatomised female torso, with a head added in pen, in a copy of the 1631 edition of Helkiah Crooke's *Mikrokosmographia* (159) (With the permission of Special Collections, University of Glasgow Library. Sp Coll Hunterian Aa.2.19. Photograph by Robert MacLean.)

readers to the uncensored image in *Mikrokosmographia*'s first edition ('*See this in the Booke at large, fol. 219*'), but for those without access to the latter, a very similar version was available in cheap, small quarto editions of Thomas Raynalde's midwifery manual *The Birth of Mankind* (Raynalde 2009: 83), first published in 1545 and reprinted into the mid-seventeenth century.

It is difficult to know how many early readers spotted the censored detail and what they thought of it, but there is evidence of intervention by users. Although the markings cannot easily be dated, in copies of the 1631 edition of *Mikrokosmographia* and the 1634 re-issue of *Somatographia* held at college libraries at the University of Cambridge, the vulva has been drawn back in.[22] Whether these marks (we might call them 'corrections') are acts of resistance to censorship or something else, they are examples of how readers of medical texts could actively participate in the material pro-duction of both the body of the book and the bodies which that book represents. A reader of a copy of the 1631 *Mikrokosmographia* in the University of Glasgow Library went one step further with the image of the female torso in the eighteenth century (or perhaps earlier), adding a pen-drawn head to the headless and limbless figure and obscuring the Table's title (Fig. 2.2).[23] The censored detail remains untouched, but the sketched head challenges the printed image's brutal cropping, making a human subject of the female body previously represented as an object of study. Her carefully shaded face, with its determined look and tightly pursed lips, gazes out of the book and into a world elsewhere, while the hair tied with a ribbon gives an origin for the printed floating lock that appears over her left shoulder, inviting contemplation of the head's absence: behind the neck, out of sight, we can imagine the nexus of print and manuscript. Censored during publication and extended by a marking reader, this image bears the material traces both of its tortured birth as an image denounced as 'indecent', and of its reception as something more.

Such examples urge larger questions about actual readership. As with most early modern printed books, it is difficult to trace patterns of owner-ship and use of medical texts. Finding hard evidence that gynaecological and obstetrical texts were treated as pornography is close to impossible, although John Cannon's admission in his manuscript memoirs that—as an adolescent at the turn of the eighteenth century—he was caught mastur-bating with a copy of Nicholas Culpeper's *Directory for Midwives* (1651) is an intriguing exception (Hitchcock 1997: 28–29).

What can be said about the reading history of *Mikrokosmographia* and *Somatographia*? Copies held in libraries in the UK, the USA and Canada contain material traces of purchase, ownership and reading practices in the early modern period. I will continue to reference actual readers' manu-script paratexts throughout the rest of this chapter, but here I offer an overview of what they suggest about the books' circulation and readership in the seventeenth and eighteenth centuries, my understanding of which is informed by Jillian Linster's doctoral work as well as my own research.[24] Bookplates and ownership inscriptions, most often found on title-pages and fly-leaves, indicate that copies were owned by not only male medical professionals (including surgeons and man-midwives),[25] but also clergy-men,[26] a ship-owner and adventurer,[27] and a lawyer.[28] There is evidence of female owners as well.[29] Inscriptions on the opening leaves to the Elham Parish Library copy of the 1631 *Mikrokosmographia* indicate it was bought by Henry Oxinden of Barham (1609–1670), a country gentleman, for 19 shillings and 6 pence on 14 June 1636, but later gifted to his second wife Katherine Oxinden née Culling (1624–1693) some time after their mar-riage in 1642: a note in his hand reads 'Katherine Oxinden her booke/ witnesse Hen: Oxinden'.[30] Another copy of the same edition has a partially effaced inscription under the summative contents page ([sec.]5r), made before the copy's 1691 entry into the University of Glasgow's library cata-logue: 'Agnes Stewart owght [i.e. aught or owns] this book' (Sp Coll Bm5-d.2). Given the various 'cultural and material practices that discour-aged women from annotating their books' (Hackel 2005: 196), female ownership and use of texts discussing women's bodies are likely to be much wider than suggested by inscriptions.

Although problems with precise dating make comparisons difficult, notations of purchase prices suggest—as we would expect—that the folio *Mikrokosmographia* was a considerable investment, and the octavo *Somatographia* was far more affordable. While a copy of the 1616 *Mikrokosmographia* was bought by John Bate—possibly the author of *The Mysteries of Nature and Art* (1635)—for 17 shillings in the early to mid-seventeenth century (similar to Henry Oxinden's payment in 1636),[31] John Babington purchased a copy of the 1634 *Somatographia* for 5 shillings, probably in the late seventeenth or early eighteenth century.[32] At least one copy of *Mikrokosmographia* was sold on privately: the back fly-leaf of a copy of the 1651 edition bears the seventeenth- or eighteenth-century inscription 'Richard Mynors me jure possidet [i.e. rightfully owns me]. /Bought of Charles Traherne/Junior of Sr Waynards parish.'[33] Beyond inscriptions of ownership and purchase, early readers marked their copies in a number of ways and for various reasons.[34] These include medi-cal notes and cross-references, corrections, dictionary work, and records of life events.[35] Printed illustrations in two copies of *Mikrokosmographia*'s first edition have been skilfully (and probably professionally) hand-coloured, although it is difficult to know when this took place.[36] Other

readers' marks, as later examples will also show, suggest a shared perception that *Mikrokosmographia* interwove the medical and the erotic. One
late seventeenth-century owner of a copy of the 1615 *Mikrokosmographia*,
Joseph Gibson, thought it appropriate to transcribe on the verso of the
text's final page (4Q3v) lines from an elegy of Ovid's scandalous *Amores*
(III.2), ending 'to the kind venus and thy boy that awes / all hearts (asist
me) I giue the my aplayse [i.e. applause]'.[37]

Gibson's inscription may seem peculiar, but its presence in a copy of
Mikrokosmographia makes more sense when we consider that discourses of
'the Ovidian body' and 'the anatomical body' were in dialogue during the
early modern period (Stanivukovic 2001: 6–7), and that women's body
parts are sometimes named after Venus in the volume (the clitoris, notes
Crooke, 'is called *aestrum Veneris* [the fire of Venus]' (238)). Most significantly, marking readers such as Gibson can be seen as responding or reacting to the fact that anatomical and midwifery manuals were rhetorically
framed in ways that provoked contemplation of the fine line between anatomical and erotic knowledge. In order to analyse this paratextual rhetoric, I turn first to *apologia*.

APOLOGIA; OR THE SHAME GAME

The modesty topos of *apologia*, a formal defence or justification, engaged
with the prevalent idea that the female reproductive organs were obscene
and shameful.[38] Crooke refers to 'the fissure that admitteth the yard' as 'a
part thought too obscoene to look vpon' (Crooke 239). The male sexual
parts were also sometimes described as 'obscene', but many medical texts
imply that female genitals are *more* obscene by observing that women's
reproductive organs are concealed by Nature, which encouraged the belief
that they were naturally shameful.[39]

Apologia, a rhetorical practice Monica Green traces back to prefaces
of late medieval gynaecological texts and the rise of the literature of
'women's secrets' (Green 2000), re-purposed female shame as authorial
shame. Early modern prefatory writers indirectly marketed texts by
expressing (or denying) regret and embarrassment for involvement in
the publication of women's 'secrets', drawing attention to their sensational exposure. Thus, the anonymous translator of a French midwifery
manual anticipates the accusation that he has offended women 'in pros-
tituting and divulging that, which they would not haue come to open

light' (Guillemeau 1612: 2¶2v–2¶3r), and in *The Birth of Mankind* the physician Thomas Raynalde responds to midwives' complaints that 'every boy and knave had of these books, reading them as openly as the tales of Robin Hood' (Raynalde 2009: 22). Although ostensibly deployed to legitimise texts and defend women's honour, such rhetoric enabled paratextual authors to fetishise the works they were introducing by analogising them with the female 'privities' they exposed, thereby appealing to illegitimate, prurient readers.

Crooke's defence was common: the impropriety of publishing women's 'secrets' was necessary if women were to help themselves or to be helped by men. His preface to Book 4 of *Mikrokosmographia* admits uncertainty about its inclusion, but goes on to declare its 'lawfull scope' and argue that fear of indecency cannot justify omission of the book because of the 'knowledge' it furnishes:

> Indeede it were to be wished that all men would come to the knowledge of these secrets with pure eyes and eares, such as they were matched with in their Creation: but shall we therefore forfet our knowledge because some men cannot conteine their lewd and inordinate affections? (197)

Here and elsewhere in *Mikrokosmographia*, 'knowledge' is a fraught term that stands for both the anatomical knowledge whose public value justifies the publication and the erotic knowledge whose private use the book potentially facilitates. Valerie Traub compellingly argues that Crooke 'simultaneously entices the reader with, and chastises him for, his desire for further knowledge', but we might question the idea that he does so in 'an effort to maintain the chastity of the readers' (2002: 112). Crooke draws attention to Book 4's inevitable abuse by implying that all readers—as fallen creatures undone by Adam and Eve's desire for knowledge—have 'lewd and inordinate affections', only some can 'conteine' them and some cannot. It follows that to read this sexually provocative book clinically is to read it unnaturally, that is, against our fallen natures. Crooke's 'pure eyes and eares' may gesture to a desirable readerly chastity, but it also aligns that chastity with man's lost innocence. In particular, 'pure eyes' resonates with other paratextual mediations of the reader's gaze. 'To gaze with Babe-like minde, can breed no blame', declares Ambrose Fisher in his tri-lingual prefatory poem, only to suggest the impossibility of such innocence through references to Adam and the 'Tree of *Knowledge*'. Whether or not we have the 'pure eares' not to hear Fisher's clever slip into the language of breeding, that he—like Jaggard—was probably blind ironises his celebration of the pure gaze.[40] One reader of a copy of *Mikrokosmographia* held at the Folger Shakespeare Library was evidently anxious about his right to look upon its contents: on the back fly-leaf one Thomas Yull has written his name repeatedly as well as 'god give him grace there on to look' (Tamara Harvey 2008: 7n7).

Such readers may have found comfort in the idea that the human body was 'an Epitome or compend [i.e. compendium] of the whole creation', but the symbolic transformation of the private body's *'Microcosme* or little worlde' (Crooke 2) into the public commodity that was *Mikrokosmographia* (or even its own epitome *Somatographia*) had the potential to heighten anxieties. Indeed, the volume repeatedly conflates book and body. The summative contents page shows that most of the thirteen books focus on different sets of body parts. Book 6, appropriately placed in the middle of the book, is "Of the Middle Region, called the Chest" (¶3r). In his preface to Book 4, Crooke claims that 'my labour would be but lame if it wanted this limbe' (197), but he also announces that Book 4 is a limb designed to be excised from the body of the volume:

> we haue so plotted our busines, that he that listeth may separate this Booke from the rest and reserue it priuately vnto himselfe (197)

Crooke casts the (male) reader as a textual surgeon, informing him that he has the power to purge the volume of this book on the generative organs. But Crooke also reminds the reader he has the right to 'reserue' the book for his private use, whatever that may entail. If a major reason why printed medical texts were perceived as pornographic was because they brought 'descriptions of the sexual body into public space, opening them indiscriminately to the gaze of many, rather than reserving them to the realm of the private and personal' (Toulalan 2007: 161), then the option of making part of a published text (a text 'made public') a private document further sensationalises the material. Separating a portion of text from an expensive, heavy folio, moreover, would not necessarily have limited its public exposure. It might even have made that material easier to circulate or sell on. Tamara Harvey observes that Crooke is 'constructing the [volume] so it ... can be modest' (2008: 7). But the suggestion that Book 4 can and perhaps should be removed from the volume arguably works to fetishise the material, provoking curiosity rather than inspiring modesty.

Although we can only speculate about their reasons for doing so, it would seem that readers did extract parts of Book 4, most notably the leaves containing chapters on "Spermatical vessels" (tubes believed to carry women's 'seed'), "the Testicles of women" (ovaries), "the Leading vessels" (Fallopian tubes) and the beginning of "the wombe or Matrix" (Chapters 10–13; V1–V4 in Crooke 1615; V2–V5 in Crooke 1631), which feature woodcuts of 'the lower belly of a woman' (Table 5), the previously discussed anatomised female torso (Tables 6 and 8) and excised female reproductive organs (Tables 7 and 9).[41] One copy of the 1631 edition lacks Books 4 and 5 in their entirety (S5-2G5), leaving a gap between Book 4's interrupted preface and the final leaf of Book 5, whose verso is Book 6's first page.[42] Such extractions are difficult to date, but it is quite possible they were performed by early readers influenced by Crooke's statement.

The state of these copies reveals that, from a bibliographical perspective, Crooke's invitation to separate Book 4 is something of a rhetorical flourish. Book 4 cannot be removed without making it obvious that the volume is defective.[43] In all editions the book is on the contents page, begins and ends in the middle of sheets,[44] and the work as a whole is continuously paginated. To remove Book 4 and/or Book 5 whole or in part, as it would appear certain readers did, is a conspicuous expurgation or—to use a contemporary term—'gelding' of the text (*OED* geld v.[1] 2b), an action that casts concern on the missing material and its use once separated. Like the censored woodcut of the female torso, whose erased part is signposted, a copy of *Mikrokosmographia* that lacks leaves on the reproductive organs announces the fetishised absent-presence of what is missing.

We need look no further than the title-page of the first edition (see Fig. 2.3), which pictures a partially-flayed woman next to a fully-flayed veined man,[45] to know that the volume includes material on the female reproductive organs. The illustration of the woman, who—covering her genitals with one hand and touching her breast with the other—holds what has been identified as the pose of a 'Venus *pudica*' or 'the modest Venus' (Traub 2002: 117–23), is also part of Book 4, where it is presented as 'the portrature of a woman great with child' (Table 10; 226).

The use of the image on the title-page may have been a marketing ploy or—as O'Malley claims (1968: 8)—an act of defiance towards the disapproving College of Physicians. Either way, the figure, who meets the reader's gaze in a way that the pen-drawn head discussed earlier refuses to, embodies the rhetorical strategy of *apologia*. For Valerie Traub, the image

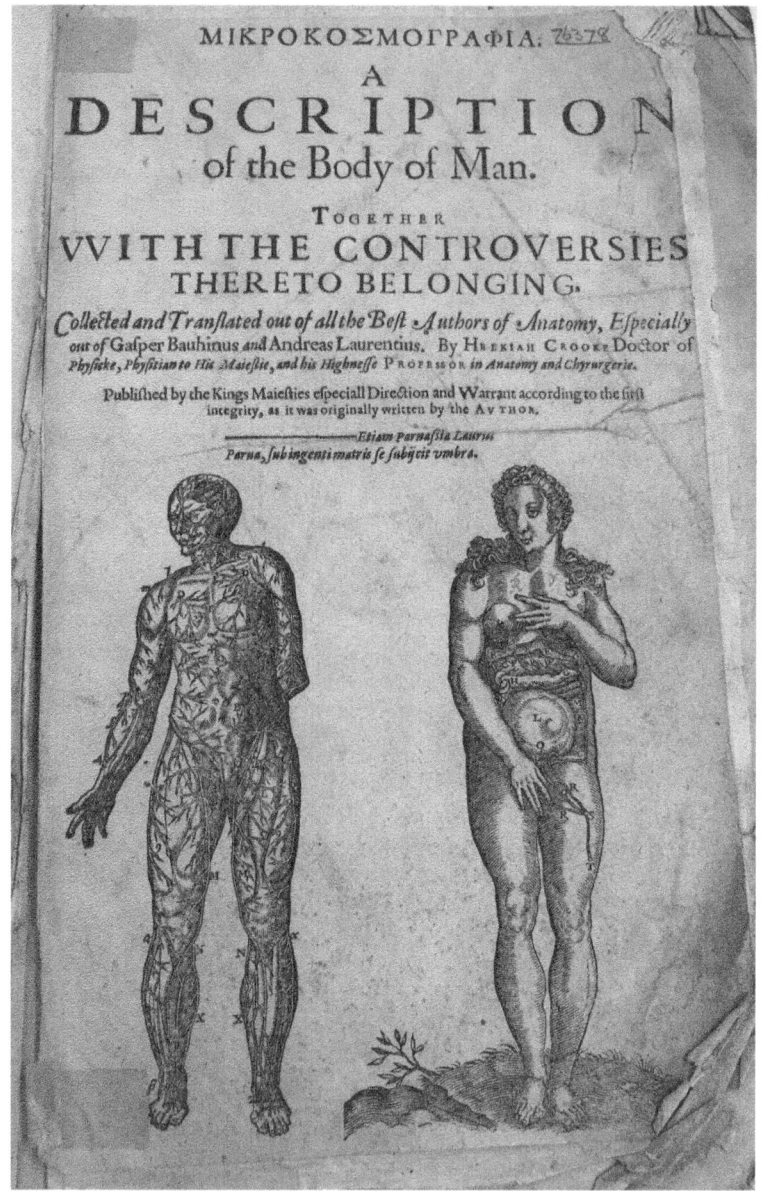

Fig. 2.3 Title-page to undated first edition of *Mikrokosmographia* (Wellcome Library, London, 1685/D). The imprint that should be at the foot of the page is missing (Photograph by Jillian Linster)

is one of many that exemplifies how the 'tensions between scientific inquiry and illicit titillation that surface in addresses to the reader are mirrored in the tension between modesty and exposure evinced in anatomical engravings of women' (2002: 112). Like *apologia*, the pictured woman enacts modesty in order to incite a curiosity that is not purely scientific, troubling one's sense of book-browsers' and owners' material interactions with the body of the book, of which the title-page's fragile leaf is a part. Perhaps responding to the printed characters that label the female figure, one early reader of a first edition chose to write his or her initials, 'R Y', above her breasts (Wellcome Library, London, 1685/D; Fig. 2.4). The inscription transforms her body into a symbol of the book, and her left hand—with its index finger pointing at the first initial—into a manicule gesturing towards the reader's ownership. Although 'R Y' will probably never be identified, the manuscript paratext testifies to early modern readers' responsiveness to the shame game of *apologia*, performed rhetorically, visually and materially by printed prefatory materials to books discussing women's bodies.

ADMONITIO, OR DO NOT SEE OVERLEAF

Since the late medieval period, *apologia* had been part of a 'doubled polemic' in prefaces to gynaecological and obstetrical texts because it was often found in conjunction with *admonitio*, a warning to or sharp rebuke of readers who might misuse the text for jocular or prurient purposes (Green 2000: 28). Although very different in tone, both served a need to market the text in the early modern book trade. While *apologia* justified the text in the eyes of legitimate readers, *admonitio* kindled the interest of supposedly illegitimate readers by making the knowledge imparted seem all the more forbidden. Sometimes *admonitio* took the form of a gentle warning, as in Crooke's anticipation of his description of 'obscoene parts' in the preface to Book 4:

> we will first describe the parts of generation belonging to men, and then proceede to those of Women also; of which wee would aduise no man to take further knowledge then shall serve for his good instruction. (199)

Although seemingly speaking in the name of modesty, Crooke again puts pressure on the meaning of the word 'knowledge', teasing out a fluidity between anatomical and erotic knowledge that would have been essen-

Fig. 2.4 Detail from title-page to undated first edition of *Mikrokosmographia* (Wellcome Library, London, 1685/D) (Photograph by Jillian Linster)

tial to the diverse appeal of *Mikrokosmographia* and especially Book 4 in the book market.

Admonitio could also be delivered as a sharp rebuke. Towards the end of the prefatory epistle to *The Expert Midwife* (1637), the small-quarto English translation of Jakob Rüff's *De conceptu et generatione hominis* (1554), an explosive denouncement of abusive readers by the anonymous author (probably the translator) follows his courteous address to welcome readers:

> To conclude, I say onely this, my intentions herein are honest and iust, and my labours I bequeath to all grave, modest and discreet women, as also to such as by profession, practise either Physicke or Chirurgery. ... But young and raw heads, Idle serving-men, prophane fidlers, scoffers, jesters, rogues; avant, pack hence; I neither meant it to you, neither is it fit for you. (A4v–A5r)

Although the reprimand's association of abusive reading with youthful and socially-transgressive idleness ('Idle serving-men' neglect their duties, and idlers become 'fidlers') may reflect genuine fears, it could serve a hidden agenda. The author of the preface probably would have known that labelling readers (or browsing book-buyers) as 'prophane fidlers'—a sexually-charged tag connoting masturbation and prostitution—and telling them to close a book because its contents are unfit for their eyes would only inspire them to read on, to turn the page and see 'the particular contents [that] follow in the next leafe' (A5r).[46]

Such admonitions, with their potential to fetishise the text and titillate the reader, rhetorically conflated the reader's action of turning the prefatory leaf to reveal the book's contents with medical authors' disclosures as they supposedly—in Crooke's words—'reveyle[d] the veyle of *Nature*' and 'ensnare[d] men's mindes by sensual demonstrations' (197). Crooke's preface to Book 4 is effectively the 'veyle' a reader must peel aside to arrive at the book which will 'reveyle' the unknown secrets he or she desires to look upon, including those parts of women 'least known' because 'veiled by Nature' (197). The hand that turns the page and manipulates the body of the book becomes an invasive one, a hand capable of defiling Nature's secrets.

In focusing on the rhetorical, visual and material qualities of printed and manuscript paratexts to early modern gynaecological and obstetrical texts, this chapter has illustrated the fetishistic functions of medical paratexts in the early modern book trade. Although the 'indecent readers' of my title are more implied than actual, the publication and reading history of *Mikrokosmographia* and its epitome suggest that early readers did participate in a meta-discourse on the relationship between medical and erotic 'knowledge', and responded to the book/body analogies posited by the volume's physical form and prefatory rhetoric. As modern readers who create and respond to our own paratexts, from library catalogue entries to the digital images and metadata of *Early English Books Online*, or even publications like this, we too have a role to play in the movement of early modern medical books as objects travelling through time and space. And while volumes such as *Mikrokosmographia* are now much more likely to speak to the archival fetish than any desire for pornographic content,[17] they continue to be framed as seductive bodies of knowledge, bibliographical repositories of early modern medical culture and practice.

Acknowledgement I would like to thank those people who generously offered advice on earlier drafts of this chapter, including Jillian Linster, Tom Lockwood, John Jowett, and Victoria Yeoman.

NOTES

1. Kassell observes that the 'production of books about women's bodies peaked in the 1650s' (2013: 66).
2. I use the term 'pornography' advisedly (see Toulalan 2007, especially 22–23). The word entered the English language some time in the mid-nineteenth century, and literally means 'writing about prostitutes' (*OED* n.). In the printed paratexts under discussion, the language of prostitution is sometimes employed to invoke the potentially pornographic nature of the texts being introduced.
3. Unless stated otherwise, references to *Mikrokosmographia* are to the first edition of 1615.
4. On the physiology of reading, see Craik (2007), and Smith (2012): 188ff. Craik focuses on pornography in Chap. 6 (115–34).
5. On the relationship between medical and erotic literature in early modern England, see: Thompson (1979: 158–175); Cressy (1997: 39–41); Sugg (2007: 112–16); and Gilbert (2002: 139–42). Sarah Toulalan (2007) discusses a range of 'medical, midwifery and quasi-medical literature' (16) in her study of pornography in seventeenth-century England. In part, my chapter builds on her observation that 'while we might assume that the primary purpose of an author of a midwifery manual was educational, we can never be sure that he or she did not deliberately word descriptions in such a way that a reader might also find them sexually arousing while at the same time disclaiming that the work had any obscene intent or function' (8). The most important work on medical texts and the early modern English book trade is Furdell (2002), although it does not address erotic representations or perceptions of books on the female body.
6. For other work on the rhetoric of prefatory materials to early modern anatomical, gynaecological and obstetrical texts, see: Traub (2002: Chap. 2 passim 77–124); Elizabeth Harvey (1992: 89–93); and Keller (2007:

Chap. 2 passim 47–70). Also see King, Chap. 1 on Spanish, Italian, German and French texts (29–64), with some consideration of English owners and users of the pan-European compendium *Gynaeciorum libri* (first published 1566). Hobby demonstrates the advantages of 'applying literary critical techniques to analyse such features as metaphor and tone' in early modern scientific texts (2008: 35).

7. My distinction between 'implied' and 'actual' readers is informed by Iser (1978). 'Legitimate' and 'illegitimate' are my own terms for the implied readers of gynaecological and obstetrical texts.

8. On the applicability of Gérard Genette's terms 'paratext', 'epitext' and 'peritext' in early modern studies, see Smith and Wilson (2011).

9. On Crooke's biography, see O'Malley (1968) and Birken (2004).

10. On Jaggard's career, see Willoughby (1934).

11. The relevant Latin manuscript annals are housed in the Royal College of Physicians, as are the transcription and translation produced in the 1950s (Book 3, 1608–29, trans J. Emberry and S. Heathcote, 1953–55, 11 November 1614, p. 65; 3 April 1615, p. 71). My account of the controversy surrounding the publication of *Mikrokosmographia* is informed by O'Malley (1968: 7–11); Russell (1963: 219); Furdell (2002: 52); and Sawday (1995: 225–6).

12. I have used the longer title of the 1615 variant issue in my Works Cited. The addition to the title runs from '*Physitian to His Maiestie*' to '*as it was originally written by the Author*'. The only other difference is the absence of London from the variant title-page (Russell 1963: 219).

13. On objections to vernacular medical texts in early modern Europe, see Traub (2002: 109–10) and Eamon (1994: 102–3).

14. See Moulton (2000: 119–57) on the 'role the mythos of Aretino played in English culture in the late sixteenth and early seventeenth centuries' (120). As Traub notes, Crooke's reference to 'Aretine' is absent from the 1631 edition (Traub 2002: 110), for which the epistle was edited and supplemented by another epistle from Crooke "To the Yovnger Sort of the Barber-Chirvrgians" (dated 30 October 1630). The omission may be related to the censorship I discuss later.

15. The volume ends with an address from Jaggard to the reader, in which he notes the venture has been 'to my no meane care and cost' (Crooke 1615: 4Q4r).

16. Like the 1631 *Mikrokosmographia*, *Somatographia*'s re-issue was bound with Ambroise Paré's *Explanation of the Fashion and Vse of Three and Fifty Instruments of Chirvrgery* (1631).

17. Linster investigates this act of censorship in her chapter on "The Censored Body" (2017: 74–101).

18. The image evidently originates from an illustration in Vesalius (1998–2002: Vol. 4, Book V, Figure 25).

19. The uncensored version appears as Table 27 of Book 3 (159/P2r) and Tables 6 and 8 of Book 4 (219/V2r; 222/V3v) in the first edition of *Mikrokosmographia* and its 1616 and 1618 re-issues. The censored image features as Table 6 for the section "Of the Naturall Parts belonging to Generation" in *Somatographia* and its re-issue (126v and 127r). It can also be found at Table 27 of Book 3 (159/P3r) and Tables 6 and 8 of Book 4 (218/V2v; 222/V4v) in the 1631 edition of *Mikrokosmographia*, and Tables 6 and 8 of Book 4 (162/P3v; 165/P5r) in the 1651 edition.

20. Jaggard, blind from 1612, is likely to have delegated many of his responsibilities, especially to his son Isaac, who by 1616 was 'a compositor and stone-man of great ingenuity' (Willoughby 1943: 116).

21. Linster argues that the 'blank space' acts as a visual representation of the vulva's status in early modern England as 'nothing' (punning on 'nothing'), and thus 'reifies' the pun's 'linguistic elision of female genitalia' (2017: 75).

22. The copy of the 1631 *Mikrokosmographia* based at Crooke's alma mater St. John's College (Ll.3.19; http://collpw-newton.lib.cam.ac.uk/vwebv/holdingsInfo?bibId=466156) was once owned by Sir Thomas Bendish (1607–1674), Ambassador to the Ottoman Empire. Its three censored images of the female torso have been marked by a reader or readers, twice in pen and once in pencil. University Library's copy of the 1634 *Somatographia* (Syn.7.63.318; http://search.lib.cam.ac.uk/?itemid=|cam brdgedb|2066924), which has a similar pencil marking, bears a late seventeenth-century inscription of ownership on the back fly-leaf: 'John […] E[x] libris[?] 1686'. (The second word, probably a surname, has been obliterated.) These examples are cited and discussed by Linster (2017: 98–101), who interprets the markings as acts of resistance which 'restore[ ] the female figure as human' (99–100).

23. University of Glasgow Library, Sp Coll Hunterian Aa.2.19; http://encore. lib.gla.ac.uk/iii/encore/record/C__Rb1606622;jsessionid=017BAB6E6 2F21ED7A2A79CB0D0EF6861?lang=eng. The earliest known owner of this copy, bound in seventeenth-century blind-tooled speckled calf, was Nicholas Downing ('Nic Downing' is inscribed on title-page), perhaps 'the London man midwife of the same name fl. 1635'. We cannot know whether Downing produced the manuscript head, but the drawing was likely produced before the Hunterian bequest of 1807. My thanks to Robert MacLean for his generous private correspondence about this and other copies of *Mikrokosmographia* housed at the University of Glasgow Library, and for his advice more generally.

24. Jillian Linster recently completed her PhD in the Department of English at the University of Iowa: *Books, Bodies, and the 'Great Labor' of Helkiah Crooke's "Mikrokosmographia"* (2017). I am grateful to Jillian for her extensive and detailed correspondence on the publication and reading his-

tory of *Mikrokosmographia* and *Somatographia*, and for allowing me to consult and cite her thesis.

25. The verso of the title-page to a copy of the 1616 *Mikrokosmographia* bears the bookplate of John Whitfield, an eighteenth-century surgeon from Shropshire (King's College London, Foyle Special Collections, St. Thomas Historical Books Collection, FOL. QM21 C76). As noted above, the earliest known owner of the Glasgow copy of the 1631 edition discussed earlier (op. cit. 23), Nicholas Downing, was possibly a man-midwife. The copy was later acquired by the famous man-midwife William Hunter (1718–83).

26. A copy of the 1616 *Mikrokosmographia* was owned by Francis Whiddon, rector at Morton Hampstead in Devon from 1617 to 1650, who appears to have given the book to John Prideaux (1578–1650), rector to Exeter College Oxford and later Bishop of Worcester from 1641 (Worcester Cathedral Library; McKeown 2014).

27. A copy of the 1615 *Mikrokosmographia* (University of Iowa, John Martin Rare Book Room, Heirs of Hippocrates No. 405) was owned by one 'Arthur Champernowne' (inscribed on the front fly-leaf in a seventeenth-century hand). The owner was probably 'Sir Arthur Champernowne', who was grandson to the sixteenth-century naval commander of the same name (Trim 2004) and 'fished at New England during 1621 and 1622' (Gray 1992: 186).

28. See note 30 for the owner who was an attorney.

29. The poet Anne Bradstreet (1612/13–1672) evidently owned or had close access to a copy of *Mikrokosmographia*, which she references and paraphrases in *The Tenth Muse*, published in 1650 (see Tamara Harvey 2008: 17–50 passim). Two different copies of the 1634 *Somatographia* had early female owners/users: an inscription at the foot of the title-page to a copy at the University of Iowa (John Martin Rare Book Room, Heirs of Hippocrates No. 459.5) reads 'Anne Hodgson booke', and a copy at the Royal College of Physicians (D2/80-g-5) has been inscribed twice by Elizabeth Proctor—'Elizabeth P' (2D1r) and 'Eth[?] Proctor' (2D8v).

30. Elham 123, held on deposit at Canterbury Cathedral Library. The copy was later bequeathed to the Canterbury surgeon John Warly (1674–1732) in the late seventeenth century, and then passed on to his son Lee Warly (1714–1807), an attorney. My thanks to Karen Brayshaw, Librarian to Canterbury Cathedral, for her advice and allowing me to access this copy. For more on Elham Parish Library, the Oxindens and the Warlys, see Hingley (2004).

31. University of Glasgow Library, Sp Coll S.M. 1935, http://encore.lib.gla. ac.uk/iii/encore/record/C__Rb3127769?lang=eng. There is an obliterated inscription at the head of the title-page: 'John Bate his booke pret. Xvij S'.

32. University of Toronto, Thomas Fisher Rare Book Library, acad 01651. The title-page bears this inscription in a late seventeenth- or early

eighteenth-century hand: 'Ex lib: Joh: Babington/pret: 5s'. This suggests *Somatographia* may have been on the expensive side, as on average unbound octavos cost between 1 and 4 shillings in the late seventeenth century (Finkelstein and McCleery 2013: 112). Focusing on the earlier period of 1550–1640, Francis R. Johnson claims 'the average illustrated book was priced 75–100 per cent higher than other books of the same number of sheets' (Johnson 1950: 90). It is unclear whether the prices given by Oxinden, Bate and Babington include binding costs.

33. University of Washington, Suzzallo and Allen Libraries, Sp Coll Rare Books, 611 C882m. 'St Waynards parish' refers to St Weonards Parish Church in Herefordshire, near the manor of Treago, which 'has been held by the Mynors family since the early fourteenth century' (Emery 2000: 586).

34. See Sherman (2008) on the diversity of early modern readers' manuscript annotations in printed books more generally.

35. The 1616 *Mikrokosmographia* once owned by John Bate, discussed earlier, has a full page of manuscript English notes on the muscles and bones of the leg with references to other parts of the text in a neat seventeenth-century hand, perhaps that of Bate himself (op. cit. 31; 806). A page in a 1616 *Somatographia* (Royal College of Physicians, D2/80-g-3; 127r) shows both a correction ('left testicle' is changed to 'right testicle') and—in a different, seventeenth-century hand—a definition for the highlighted word 'excoriated' ('skin flead [i.e. flayed] or fretted off'), indicating probable consultation of Cotgrave 1611 (2M5v) or Blount 1656 (Q2r), perhaps in later editions. The verso of the final leaf of the text proper in a copy of the 1615 *Mikrokosmographia* bears a series of inscriptions recording dates and times of births and deaths in a man's immediate family between 1763 and 1765 (Royal College of Surgeons, uncatalogued; 4Q3v).

36. Royal College of Surgeons, uncatalogued; British Library, General Reference Collection, 781.k.1. Both are first editions, but whereas the RCS copy was definitely published in 1615, the BL copy cannot be dated, as it lacks an original title-page.

37. The copy resides in the Crerar Rare Books Room at the University of Chicago (f QM21.C9 1615; https://catalog.lib.uchicago.edu/vufind/Record/1257383). The transcribed lines, which seem to draw on a translation that appears in Sandys (1615: 298, reprinted into the 1630s), are followed by an ownership inscription in the same hand: 'Joseph Gibson / his booke / 1694'.

38. On the shame associated with the female body in the early modern period, see Paster (1993), especially Chap. 4 (163–214).

39. This notion was strengthened through expressions of the Galenic one-sex model. 'That which man hath apparent without, that women have hid within' wrote the French surgeon Ambroise Paré (1634: 128). See Kassell (2013: 63) on Crooke's inconsistent position on the one-sex model.

40. Ambrose Fisher was probably the 'blind man of letters and erudite dramatist' of the same name (d. 1617), discussed by Cuttica (2012: 28–29).
41. The 1616 *Mikrokosmographia* at the Royal College of Physicians (D1/37-a-16) is missing leaves V1–V2 (now replaced by facsimiles), and a copy of the first edition at the Wellcome Library (1685/D; the date is missing from the title-page) lacks V2–V4. The 1631 editions of Elham Parish Library (op. cit. 30), New College Library, Oxford (BT3.237.8) and Trinity College Library, Oxford (O.9.17) are missing V3–V4, V4 and V5 respectively. The latter copy, which has an inscription of donation to Trinity College in 1644 at the foot of the engraved title-page, also lacks leaves X1 (featuring Tables 11 and 12, which show images of a womb, foetus and after-birth) and X6 (Chap. 16 on 'the Lap or Priuitie'). (Jillian Linster, private correspondence).
42. This copy was sold on eBay on 16 October 2007 along with another copy of the same edition and a copy of the 1651 edition ("1631 Crooke ~ Mikrokosmographia ~ English Medical Folio", http://www.worthpoint. com/worthopedia/1631-crooke-mikrokosmographia-english-medical).
43. My thanks to John Jowett for drawing this to my attention.
44. Book 4 runs from S3r–Z2v in the first edition and its re-issues, and from S4r–Z3v in the 1631 edition. Its placement in the 1651 edition (O1v–R5r) makes its removal even more difficult because it shares a leaf with the end of Book 3 and finishes on the same page on which Book 5 begins.
45. The images of both the man and the woman are probably 'borrowed from Bauhin's *Theatrum anatomicum*' (O'Malley 1968: 8).
46. For the sexual meanings of the noun and verb 'fiddle' and the noun 'fiddler', see Williams (1994: I, 478–80).
47. On the archive as fetish, see LaCapra (1985: 20, 92).

Bibliography

Primary Sources

Blount, Thomas. 1656. *Glossographia: Or A Dictionary.* London. Print.
Cotgrave, Randle. 1611. *A Dictionary of the French and English Tongues.* London. Print.
Crooke, Helkiah. 1615. *Mikrokosmographia* [Greek]. *A Description of the Body of Man. Together with the Controuersies Thereto Belonging. Collected and Translated Out of All the Best Authors of Anatomy, Especially Out of Gasper Bauhinus and Andreas Laurentius. By Helkiah Crooke Doctor of Physicke, Physitian to his Maiestie, and his Highnesse Professor in Anatomy and Chyrurgerie. Published by the Kings Maiesties especiall Direction and Warrant according to the first integrity, as it was originally written by the Author.* London. Re-issued 1616 and 1618. Print.

———. 1616. *Somatographia Anthropine* [Greek]. *Or A Description of the Body of Man. By Artificiall Figures Representing the Members, and Fit Termes Expressing the Same. Set Forth Either to Pleasure or to Profite Those Who Are Addicted to This Study. By W.I. Printer.* London. Re-issued 1634. Print.

———. 1631. *Mikrokosmographia* [Greek]. *A Description of the Body of Man.* 2nd ed. London. Print.

———. 1651. *Mikrokosmographia* [Greek]. *A Description of the Body of Man.* 3rd ed. London. Print.

Guillemeau, Jacques. 1612. *Child-Birth or, The Happy Deliuerie of Women.* Trans. anon. London. Print.

Norton, Thomas, and Thomas Sackville. 1570. *The Tragidie of Ferrex and Porrex.* London. Print.

Paré, Ambroise. 1634. *The Workes of That Famous Chirurgion Ambrose Parey Translated Out of Latine and Compared with the French. by Th: Johnson.* London. Print.

Raynalde, Thomas. 2009. *The Birth of Mankind: Otherwise Named, The Woman's Book,* ed. Elaine Hobby. Farnham: Ashgate. Print.

Rüff, Jakob. 1637. *The Expert Midwife, or an Excellent and Most Necessary Treatise of the Generation and Birth of Man.* Trans. anon. London. Print.

Sandys, George. 1615. *A Relation of a Journey Begun An: Dom: 1610.* London. Print.

Vesalius, Andreas. 1998–2002. *On the Fabric of the Human Body.* Trans. William Frank Richardson and John Burd Carman. 5 vols. San Francisco: Norman. Print.

Secondary Sources

Birken, William. 2004. Crooke, Helkiah (1576–1648). *Oxford Dictionary of National Biography.* Oxford University Press. http://www.oxforddnb.com/view/article/6775. Accessed 9 Sept 2016.

Camille, Michael. 1997. The Book as Flesh and Fetish in Richard de Bury's *Philobiblon.* In *The Book and the Body,* ed. Dolores Warwick Frese and Katherine O'Brien O'Keeffe, 34–77. Notre Dame: University of Notre Dame Press.

Craik, Katharine. 2007. *Reading Sensations in Early Modern England.* Basingstoke: Palgrave Macmillan.

Cressy, David. 1997. *Birth, Marriage, and Death: Ritual, Religion, and the Life-Cycle in Tudor and Stuart England.* Oxford: Oxford University Press.

Cuttica, Cesare. 2012. *Sir Robert Filmer (1588–1653) and the Patriotic Monarch: Patriarchalism in Seventeenth-Century Thought.* Manchester: Manchester University Press.

Eamon, William. 1994. *Science and the Secrets of Nature: Books of Secrets in Medieval and Early Modern Culture.* Princeton: Princeton University Press.

Emery, Anthony. 2000. *Greater Medieval Houses of England and Wales: 1300–1500. Vol. II: East Anglia, Central England and Wales.* Cambridge: Cambridge University Press.

Finkelstein, David, and Alistair McCleery. 2013. *An Introduction to Book History.* 2nd ed. London: Routledge.

Furdell, Elizabeth Lane. 2002. *Publishing and Medicine in Early Modern England.* Woodbridge: Boydell & Brewer.

Gilbert, Ruth. 2002. *Early Modern Hermaphrodites: Sex and Other Stories.* New York: Palgrave.

Gray, Todd. 1992. Fishing and the Commercial World of Early Stuart Dartmouth. In *Tudor and Stuart Devon: The Common Estate and Government*, ed. Todd Gray, Margery Rowe, and Audrey Erskine, 173–199. Exeter: University of Exeter Press.

Green, Monica Helen. 2000. From 'Diseases of Women' to 'Secrets of Women': The Transformation of Gynaecological Literature in the Late Middle Ages. *Journal of Medieval and Early Modern Studies* 30 (1): 5–39.

Hackel, Heidi Brayman. 2005. *Reading Material in Early Modern England: Print, Gender, and Literacy.* Cambridge: Cambridge University Press.

Harvey, Elizabeth D. 1992. *Ventriloquized Voices: Feminist Theory and English Renaissance Texts.* London: Routledge.

Harvey, Tamara. 2008. *Figuring Modesty in Feminist Discourse Across the Americas, 1633–1700.* Aldershot: Ashgate.

Hingley, Sheila. 2004. *The Oxindens, Warlys and Elham Parish Library: A Family Library and Its Place in Print Culture in East Kent.* PhD Thesis, Canterbury Christ Church University College, 2 vols.

Hitchcock, Tim. 1997. *English Sexualities, 1700–1800.* London: Macmillan.

Hobby, Elaine. 2008. 'Dreams and plain dotage': The Value of *The Birth of Mankind* (1540–1654). In *Literature and Science*, ed. Sharon Ruston, 35–52. Cambridge: D. S. Brewer.

Iser, Wolfgang. 1978. *The Act of Reading: A Theory of Aesthetic Response.* Baltimore: John Hopkins University Press.

Johnson, Francis R. 1950. Notes on English Retail Book-Prices, 1550–1640. *The Library* 5 (2): 83–112.

Kassell, Lauren. 2013. Medical Understandings of the Body, c. 1500–1750. In *The Routledge History of Sex and the Body: 1500 to the Present*, ed. Sarah Toulalan and Kate Fisher, 57–74. New York: Routledge.

Keller, Eve. 2007. *Generating Bodies and Gendered Selves: The Rhetoric of Reproduction in Early Modern England.* Seattle: University of Washington Press.

King, Helen. 2007. *Midwifery, Obstetrics and the Rise of Gynaecology: The Uses of a Sixteenth-Century Compendium.* Aldershot: Ashgate.

LaCapra, Dominick. 1985. *History and Criticism.* Ithaca: Cornell University Press.

Linster, Jillian. 2017. *Books, Bodies, and the 'Great Labor' of Helkiah Crooke's "Mikrokosmographia"*. PhD Thesis, University of Iowa.

McKeown, Deidre. 2014. Researching the Provenance of Seventeenth Century Anatomy Books. Blog-post, June 13. http://worcestercathedrallibrary.blogspot.co.uk/2014/06/researching-provenance-of-seventeenth.html. Accessed 2 Sept 2016.

Moulton, Ian. 2000. *Before Pornography: Erotic Writing in Early Modern England*. Oxford: Oxford University Press.

O'Malley, C.D. 1968. The Fielding H. Garrison Lecture: Helkiah Crooke, M. D., F. R. C. P., 1576–1648. *Bulletin of the History of Medicine* 42 (1): 1–18.

Paster, Gail Kern. 1993. *The Body Embarrassed: Drama and the Disciplines of Shame in Early Modern England*. Ithaca/New York: Cornell University Press.

Russell, Kenneth Fitzpatrick. 1963. *British Anatomy, 1525–1800: A Bibliography*. Parkville: Melbourne University Press.

Saenger, Michael. 2006. *The Commodification of Textual Engagements in the English Renaissance*. Aldershot: Ashgate.

Sawday, Jonathan. 1995. *The Body Emblazoned: Dissection and the Human Body in Renaissance Culture*. London: Routledge.

Sherman, William H. 2008. *Used Books: Marking Readers in Renaissance England*. Philadelphia: University of Pennsylvania Press.

Smith, Helen. 2012. *'Grossly Material Things': Women and Book Production in Early Modern England*. Oxford: Oxford University Press.

Smith, Helen, and Louise Wilson. 2011. Introduction. In *Renaissance Paratexts*, ed. H. Smith and L. Wilson, 1–14. Cambridge: Cambridge University Press.

Stanivukovic, Goran V. 2001. Introduction: Ovid and the Renaissance Body. In *Ovid and the Renaissance Body*, ed. Goran V. Stanivukovic, 3–18. Toronto: University of Toronto Press.

Sugg, Richard. 2007. *Murder After Death: Literature and Anatomy in Early Modern England*. Ithaca, NY: Cornell University Press.

Thompson, Roger. 1979. *Unfit for Modest Ears: A Study of Pornographic, Obscene and Bawdy Works Written in England in the Second Half of the Seventeenth Century*. London: Macmillan.

Toulalan, Sarah. 2007. *Imagining Sex: Pornography and Bodies in Seventeenth-Century England*. Oxford: Oxford University Press.

Traub, Valerie. 2002. *The Renaissance of Lesbianism in Early Modern England*. Cambridge: Cambridge University Press.

Trim, D.J.B. 2004. Champernowne, Sir Arthur (*b.* in or before 1524, *d.* 1578). *Oxford Dictionary of National Biography*. Oxford University Press. Online ed., May 2009. http://www.oxforddnb.com/view/article/71675. Accessed 14 Sept 2016.

Wall, Wendy. 1993. Prefatorial Disclosures: 'Violent Enlargement' and the Voyeuristic Text. In *The Imprint of Gender: Authorship and Publication in the English Renaissance*, 169–226. Ithaca/New York: Cornell University Press.

Williams, Gordon. 1994. *A Dictionary of Sexual Language and Imagery in Shakespearean and Stuart Literature*. 3 Vols. London: The Athlone Press.

Willoughby, Edwin. 1934. *A Printer of Shakespeare: The Books and Times of William Jaggard*. London: Philip Allan.

CHAPTER 3

Touching Twins in the Texts and Medical Paratexts of Seventeenth-Century Midwifery Books

Louise Powell

The seventeenth century saw a sharp increase in the publication of midwifery books that were written or translated into English, and a corresponding rise in the number of medical references to twins. Although twins were not mentioned by every midwifery book, many works devoted chapters or a significant number of paragraphs to how to detect or deliver them. Some midwifery texts, such as Jacques Guillemeau's *Child-birth, or the Happy Deliverie of Women* (1612), and Thomas Chamberlayne's *The Compleat Midwife's Practice* (1656), also supplemented their treatments of twins with illustrations of them *in utero*. This combination of text and medical paratext reflects the didactic purpose of midwifery books, but it also reveals some broader seventeenth-century ideas surrounding twins. Whilst the written twins are always shown to be apart from each other, their illustrated counterparts are always shown to be touching. Such differences between texts and medical paratexts suggest uncertainty as to

L. Powell (✉)
Sheffield Hallam University, Sheffield, UK

© The Author(s) 2018   43
H. C. Tweed, D. G. Scott (eds.), *Medical Paratexts from Medieval to Modern*, Palgrave Studies in Literature, Science and Medicine,
https://doi.org/10.1007/978-3-319-73426-2_3

when the twin relationship actually begins, and a subsequent anxiety over the types of interaction which these figures experienced within such a confined space as the womb. Through an exploration of how Guillemeau's and Chamberlayne's works write about and illustrate twins, this chapter will argue that there was concern as to how effective and affective seventeenth-century medical practices surrounding them could be. There was an uncomfortable awareness that many of the medical ideas regarding twins were ultimately speculative in nature, and so had the potential to harm them or their relationship with each other. As these insights cannot be gathered through the analysis of words or illustrations alone, this chapter demonstrates that when texts and medical paratexts are given equal weighting, readers can uncover a new and complex set of issues.

Much of the current work featuring medical illustrations employs them as visual representations of obsolete practices. Lauren Dundes, for example, uses illustrations from a number of different eras in order to demonstrate how Western cultures have gradually shifted from a vertical birthing position to a horizontal one (Dundes 1987: 636–41), whilst Praveen Kumar Goyal and Andrew N. Williams supplement their discussion of Ambroise Paré's paediatric medical practices with visual representations of various people and instruments (Goyal and Williams 2010: 108–114). This use of medical illustrations is highly effective in such a context, as it allows the reader to gain a greater understanding of the topics being discussed. Yet I believe that medical illustrations can also play more than a supplementary role in criticism, as they often served an important paratextual function within the original works.

Early modern paratexts have been the focus of an emerging set of scholarship in literary studies, and critics have tended to examine such written varieties as dedicatory epistles and prefaces because they maintain that these features reveal insights about the author that are unavailable elsewhere. David M. Bergeron, for example, argues that 'Nowhere do we hear the author's voice more clearly or directly than in prefatory material' (Bergeron 2006: 1), whilst Helen Smith and Louise Wilson suggest that 'the explicit instructions of the preface or dedication' make it 'easier to extract an author's apparent design' (Smith and Wilson 2011: 6). This approach may have been fruitful for literary studies, but I believe that it is also somewhat problematic, because it risks privileging paratext over text, and losing sight of the relationship between them. Yet as Monica H. Green argues, medical paratexts historically informed the texts they accompanied, and vice versa: for 'later medieval Europe [...] text and image could work together dynamically to allow "double readings" – textual and visual – both subtle and literal' (Green 2009: 150). Rather than isolating medical illustrations, then, this chapter will situate texts and medical paratexts alongside each other, and discuss them together.

The medical illustrations which feature in Jacques Guillemeau's *Childbirth, or the Happy Deliverie of Women* and Thomas Chamberlayne's *The Compleat Midwife's Practice* show the various positions which children could occupy within the womb. They therefore have their origins in Muscio's *Gynaecology* (c. 500–600), a work composed in North Africa which contained a number of foetal images of single as well as twin births. These illustrations proved so popular that by the medieval period, as Green recalls, they 'circulate[d] independently of the full *Gynaecology*' (Green 2009: 151). Green stresses that because these images appeared in manuscript form, and there was a somewhat vexed relationship between midwife and surgeon, it is unclear exactly who actually viewed them, and it is likely that the audience for them was relatively small. She speculates: 'Perhaps we can imagine that the surgeons or physicians who owned these manuscripts showed the birth images to midwives with whom they worked' (Green 2009: 152).

Observations made by Thomas G. Benedek reveal that the relations between midwives and surgeons were still strained by the seventeenth century (see Benedek 1977: 564), but the fact that foetal images began to be inserted into midwifery manuals suggests that there was a larger audience for them than during the medieval period. Indeed, the works themselves offer a number of insights into how the illustrations were intended to be used. In Guillemeau's work, medical illustrations appear individually underneath chapter headings, whilst in Chamberlayne's book they are numbered and all situated together in a grid-like formation which spans two pages. These layouts both suggest that the medical paratexts were designed to supplement the texts they appeared alongside, and so they had a didactic purpose. The images would stand out from pages of text, meaning that any reader could quickly find the correct area of the book. The individual illustrations of Guillemeau would then help the reader to locate and peruse the relevant chapter, whilst the numbered images found in Chamberlayne corresponded to a marginal gloss that identified pertinent parts of a longer chapter. In this way, texts and medical paratexts combined to allow the information about particular foetal positions to be read and followed with the utmost efficiency – an important benefit, given the grim mortality rate of early modern childbirth.

If the texts and medical paratexts of midwifery manuals were designed to be used in conjunction with one another, it is not unreasonable to expect that their content would also align. Yet whilst such similarity is certainly apparent in Guillemeau's and Chamberlayne's written accounts and visual representations of single births, their treatments of twin births reveal an important number of differences. The texts maintain that there is no interaction between twins of separate bodies *in utero*, but their paratextual counterparts always show them to be touching. This mismatch may seem only to reveal the fact that the writers or translators and illustrators of the works were different people who may never have met, but as I will now argue, they are actually indicative of a number of conflicting seventeenth-century ideas surrounding twins.

The quality of touch was considered to be an important diagnostic tool for the identification of twins *in utero*. It was held to be one of the key ways in which a pregnant woman could establish that she was carrying two children, as opposed to one. According to Guillemeau and Chamberlayne, the presence of twins became evident when they both touched the inside of their mother's womb, 'on the right and the left side, at the same instant' (Guillemeau sig. B3^v; Chamberlayne sig. D8^r). This interaction with the inside of the mother's body also had the potential to make an impact upon the outside of it, as the movement could create visible marks on her abdomen. A large indentation, which looked like 'a line' that could stretch 'from the navel to the groine' (Chamberlayne sig. E1^v), was believed to emerge as a consequence of the contact between children and mother. In so doing, it offered a new, physical sign of the presence of twins.

If twins were considered able to touch their mother's body, then it should follow logically that they could interact with each other, but such contact was not associated with all who shared the same birth. In addition to offering evidence as to the existence of two children *in utero*, the quality of touch also served another diagnostic function prior to birth. As Guillemeau's and Chamberlayne's texts demonstrate, the degree of contact between twins was believed to reveal their corporeal status. Guillemeau suggests that a physician 'must observe whether the two children be monsters or no [...] which he may easily perceive, by sliding his right hand open, betweene the two heads, putting it as high as he can, to feele the division' (Guillemeau sig. Y2^r). Similarly, Chamberlayne instructs the midwife 'to try whether it be not some monstrous conception [...] which may be known, if she put up her hand gently betweene the two heads as high as she can' (Chamberlayne sig. I8ʳ). Twins who were touching, then, were believed to be conjoined, whilst those who were not interacting were considered to inhabit separate bodies. The midwife or surgeon may have been instructed to begin by establishing a gap between the twins' heads, but they were expected to ensure that there was distance surrounding their entire bodies before they could confirm whether they shared one body, or occupied two.

This emphasis upon contact seems rooted in the fact that conjoined twins share some body parts, and their separate-bodied counterparts do not. Guillemeau and Chamberlayne issue further instructions to the midwife or surgeon who is delivering twins of separate bodies, and the theme of a lack of interaction between them becomes more overt. By the time that the mother's labour has advanced, the quality of touch shifts from a diagnostic tool to something that must be avoided at all costs. The type of touch that is to be avoided depends upon whether the twins are positioned head-first, feet-first, or a mixture of the two. If they are head-first, then one twin must be guided to the birth canal, but the other must be impeded from approaching it: 'by all means bring forward, the former that he would receive, holding down the other with two or three fingers of his left hand' (Guillemeau sig. Y2ʳ). If one twin is head-first and the other feet-first, then their proximity to the birth canal dictates the order of their birth: 'if the head of the one be less forward than the feet of the second, it will be most convenient to draw that forth by the feet' (Chamberlayne sig. I8ᵛ). The child who is positioned head-first must then be moved away to allow their sibling a clearer passage: 'turn [...] the head of the other a little to the other side' (Chamberlayne sig. I8ᵛ). For twins who are both positioned feet-first, however, the quality of touch becomes potentially life-threatening. The reader is instructed to locate the feet of one twin, and to direct them gently towards the birth canal, but Guillemeau stresses just how important it is that the midwife or surgeon does not mistake one twin's foot or leg for their sibling's: 'If he should do so, then without doubt in drawing of them forth, he would tear them both asunder' (Guillemeau sig. Y4ᵛ). There is no suggestion that the twins themselves would touch, only that the surgeon or midwife could make them do so. Such forced contact nevertheless has the potential to cause both children and their mother to die, or at the very least to experience horrific bodily pain.

Guillemeau's and Chamberlayne's texts thus suggest that twins of separate bodies had the inclination to interact with their mother, but never with each other. Indeed, the midwives and surgeons seem to be encouraged to act as though they are delivering two completely different children who just happen to inhabit the same womb. This attitude does make some medical sense, because it allows whoever is delivering the twins to adopt a strictly practical and unemotional stance towards them, but it is highly problematic in other ways. If the twins are suggested never to interact with each other while they are in the womb, then their relationship loses some of its great potential for closeness. Moreover, the work of Valerie A. Fildes makes clear that newly-born twins would frequently be separated, with one sent to a wet nurse who sometimes lived a great distance away, because it was believed that the mother would only have enough milk for one child (see Fildes 1986: 77). No uterine interaction between twins would mean that their relationship would have to begin after birth, but the degree of affection which could develop if one twin was sent away while still a newborn would surely be minimal. Considered in relation to seventeenth-century beliefs and practices regarding infant nursing, Guillemeau's and Chamberlayne's texts suggest a twin relationship which is so distant that it almost cannot be termed so. The lack of interaction between such children seems to indicate a lack of interest and emotion on both sides.

Yet the very opposite sense emerges from the paratexts of these works, because here, the uterine twins are always shown to be touching. The types of interaction which take place between them, as well as small illustrative details, point towards an understanding of the twin relationship that is far more complex than the one suggested by the texts. The illustrators of Guillemeau's and Chamberlayne's paratextual foetal images are unknown, but while there is a definite degree of similarity between them, their styles seem distinctive enough to certify them as different people. Guillemeau's work contains two medical illustrations of twins, and Chamberlayne's features three, but it will be most effective to discuss them by the positions they depict, as opposed to the texts in which they appear.

Both works contain paratextual illustrations of twins who occupy opposing positions in the womb: one is head-first, and the other is feet-first. An initial glance at the two images suggests a remarkable degree of similarity between the poses of the twins as represented in Guillemeau, on the left (Fig. 3.1), and Chamberlayne, on the right (Fig. 3.2). Moreover, the twins themselves appear to be doing the same thing, because each is gripping the other's ankle with their right hand. Yet upon closer inspection, the twins within each illustration are acting differently towards their sib-

Fig. 3.1 (Left) Illustration of twins *in utero*, one positioned head-first, one positioned feet-first (From Guillemeau's *Child-birth, or the Happy Deliverie of Women* (1612), sig. Y1ʳ)

ling. In Guillemeau's foetal image, the twin who is feet-first stretches his arm fully out to grasp his brother by the ankle, but his sibling has to bend his arm upwards at an uncomfortable-looking angle in order to reciprocate the movement. The same observation also applies to the illustration which appears in Chamberlayne, and there is a further difference. Whereas the twin who is head-first is looking directly at his sibling, his brother seems to be averting his gaze entirely, and instead looking at the viewer.

The other three foetal images of twins which appear in Guillemeau's and Chamberlayne's texts also evoke an impression of similarity which is offset by small differences. All of these illustrations depict twins who are touching, and they are holding onto each other so tightly that one of each twin's arms is obscured by their sibling's body. In the two images of the feet-first twins (Figs. 3.3 and 3.4), the similar poses are offset by the fact that

Fig. 3.2 (Right)
Illustration of twins *in
utero*, one positioned
head-first, one
positioned feet-first
(From a 1697 reprint of
Chamberlayne's *The
Compleat Midwife's
Practice* (1656), sig. I6ʳ)

one twin is looking closely at their sibling, who does not return their gaze. The illustration from Guillemeau, which appears on the left, also shows that only one twin has their hand on their sibling's shoulder. This gesture also remains unreturned in the foetal image of the feet-first twins (Fig. 3.5); although the shading is not entirely distinctive, the boy on the right side of the womb has his sibling's hand on his shoulder. In addition, neither of these twins are looking at each other; instead, they appear to be gazing directly at the viewer with heads tilted at slightly different angles.

It is probable that the differences outlined in these illustrations were partly a consequence of issues of space, or a conscious artistic decision to make the images seem more interesting by ensuring that they were not identical. Yet it appears that these medical paratexts were also informed by a sense of uncertainty as to how similar or different twins could, and should, be. By unsettling the immediate impression of likeness that they

Fig. 3.3 (Left) Illustration of twins *in utero*, positioned feet-first (From Guillemeau's *Child-birth, or the Happy Deliverie of Women* (1612), sig. Y3ʳ)

Fig. 3.4 (Middle) Illustration of twins *in utero*, positioned feet-first (From a 1697 reprint of Chamberlayne's *The Compleat Midwife's Practice* (1656), sig. I6ʳ)

Fig. 3.5 (Right)
Illustrations of twins *in
utero*, positioned
head-first (From a 1697
reprint of
Chamberlayne's *The
Compleat Midwife's
Practice* (1656), sig. I6ʳ)

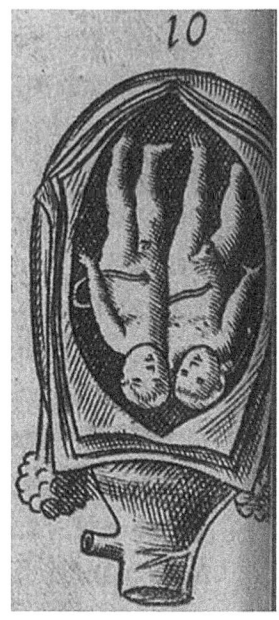

elicited, these illustrations suggest a belief that twins should be absolutely
the same as each other, and an awareness that this ideal is unrealistic.

Throughout these paratextual images of twins, there is a continual
tension between the opposing ideas of unity and division. The twins who
extend their arms fully to hold onto their siblings' ankles generate the
uncomfortable impression of a struggle for absolute dominance, while
those who hold onto their brothers' shoulders but receive no affectionate
touch in return create the impression of a desperate desire for closeness.
Then there is the fact that not one of these illustrations shows both twins
looking at each other: one of them is usually looking away, but sometimes
both of them are. There is not one image which suggests a totally unified
relationship, and none of them indicates an utterly divided one; instead,
both elements are present in all of these illustrations. What emerges from
these paratextual representations of two children who inhabit the same
womb, then, is a relationship which is not only complex for other people
to understand, but also for the twins themselves to experience.

Summarised in this way, Guillemeau's and Chamberlayne's paratextual
foetal images seem to directly contradict the impressions generated by

their textual discussion of twins. The paratexts clearly depict twins of separate bodies who touch each other, a type of interaction which the texts had limited to conjoined twins alone. The twins of the foetal images are active participants in a complicated relationship whilst they are still in the womb, but their textual counterparts show no inclination to engage with each other before they are born. What is more, the illustrated twins seem to have a connection with each other that is completely missing from the texts. When faced with evidence of such different ideas surrounding twins, the instinctive response is to feel that either the text or the paratext must be 'wrong', and to try and establish which of them is 'right'. But because texts and medical paratexts were structured to work together, I do not feel that this would be the correct approach. Rather, it seems more logical to view these two different sets of ideas together, and to interrogate exactly what they reveal about seventeenth-century attitudes towards twins.

Both the texts and medical paratexts are centred by necessity around the womb, and they each try to present what happens inside of it when its occupants are twins. In an age well before ultrasound, these attempts at representation were ultimately quite speculative, so the disparities between words and illustrations seem, on one level, to indicate the uncertainty of medical practices surrounding children who were born at the same time. If the texts suggest that twins of separate bodies do not interact but the paratexts reveal that they do, the success rate of the diagnostic tool of touch is drawn into question. As the surgeon or midwife is making their diagnosis by feeling, not seeing, there is the possibility that they may confuse conjoined twins for those who inhabit separate bodies, or vice versa. Twins who have their arms around each other, as in the 'feet-first' examples, could be misunderstood as conjoined because it would be very difficult to establish distance between them. As touch was the only diagnostic tool which surgeons and midwives had in relation to twins *in utero*, a discussion of its potential unreliability would prove both distracting and confusing. The texts may stress that twins of different bodies do not touch, then, but their paratexts function as a visual reminder that this rule might not always apply. Taken together, the words and their illustrations express both certainty and doubt as to how accurately a midwife or surgeon can establish the corporeal status of the twins they are delivering.

Whilst one explanation behind the differences between texts and paratexts can be located in an awareness of the limitations of medical practice, another is rooted in broader ideas and beliefs surrounding twins. Since the words and illustrations do not agree upon whether twins of separate bodies interact before birth, there seems to be some confusion and anxiety as to exactly when and where the twin relationship begins. Whereas the paratexts suggest that it starts to develop before birth, the texts assert that it happens afterwards, but neither answer is entirely convincing. If twins do touch and begin to negotiate their attitudes towards each other *in utero*, two problematic questions arise: what is the nature of such interaction, and what effect would their post-birth separation for nursing purposes have upon that relationship? No answer can be readily supplied, so the textual assertion that twins do not interact before birth seems more appealing. Yet this viewpoint is not without its issues either, for if twins are parted just after birth to be nursed, when exactly would they have the time to develop a relationship with each other? The distance emphasised by the texts and the closeness stressed through the paratexts thus become attempts to reconcile the necessary separation of newborn twins with the sense that their shared uterine confinement and almost simultaneous birth must create some bond between them. Both scenarios are allowed to remain, then, because neither of them completely solves the problem.

Read or viewed separately, the texts and medical paratexts of Guillemeau and Chamberlayne that address the topic of twins seem only to be concerned with how to identify them *in utero* and ensure their safe delivery. When they are combined, however, they actually point towards broader issues surrounding seventeenth-century twins. There are two main concerns arising from the combination of words and images: when the twin relationship begins, and the extent to which it is affected by a separation that is necessary for survival. Neither of these issues may be referred to directly by text or paratext, but these features do suggest two ways of reacting to them. The first of these would be to assume that as twins do not interact within the womb, their relationship would not be at all affected by a parting, and the second is to suggest that because twins do interact within the womb, the bond which they have with each other by the time that they are born is already strong enough to withstand a separation. In this way, two representations of what seem to be a solely medical nature actually come to address emotional and interpersonal aspects regarding twins as well.

As this chapter has sought to demonstrate, then, both the texts and medical paratexts of the midwifery manuals of Guillemeau and Chamberlayne offer insights into some of the practices, ideas, and concerns which surrounded twins during the seventeenth century. The quality of touch was believed to be vital for detecting the presence of twins *in utero.* Yet whilst the written chapters maintain that only conjoined twins were able to interact with each other, the illustrations depict two children who were born at the same time but inhabit separate bodies as touching. Such a discrepancy between the texts and the medical paratexts then points towards uncertainty surrounding the effectiveness of seventeenth-century medical beliefs regarding twins, as well as concerns as to whether the relationship between such children would be adversely affected when they were separated to be nursed. These medical and emotional anxieties suggest that there was a seventeenth-century discomfort at the separation of newborn twins for nursing purposes, but an awareness that such a practice was necessary if the twins were to survive. The different attitudes demonstrated within the texts and foetal images thus seem to conceptualise the twin relationship as so distant as not to exist, or so close as to overcome all obstacles, so that any emotional discomfort at their parting did not prevent it from taking place.

Yet the texts and medical paratexts of Guillemeau and Chamberlayne do not only reveal seventeenth-century understandings of, and anxieties surrounding, twins: they also indicate how these two features can inform each other. The critical tendency to isolate the paratext from the text has produced many informative works of scholarship, but such an approach would not have been particularly illuminating with the material used in this chapter. Viewed separately, the foetal images seem only to supplement the medical text, and have a solely didactic function. When texts and medical paratexts are viewed together and their differences are considered in relation to each other, however, they have the potential to transcend the topic they discuss. By being given equal weighting, the words and illustrations which seem only to address the practical issue of how to identify and deliver twins can also be seen to engage with concerns regarding the effectiveness of medical practices, and how actions which are necessary for physical wellbeing can impact upon emotional health, too. When text and medical paratext are analysed together, then, a much broader and more complex range of insights emerges than if they had been examined in isolation from each other.

BIBLIOGRAPHY

Benedek, Thomas G. 1977. The Changing Relationship Between Midwives and Physicians During the Renaissance. *Bulletin of the History of Medicine* 51 (4): 550–564.

Bergeron, David M. 2006. *Textual Patronage in English Drama, 1570–1640*. Aldershot: Ashgate.

Chamberlayne, Thomas. 1656. *The Compleat Midwife's Practice*. London: Nathaniel Brooke.

Dundes, Lauren. 1987. The Evolution of Maternal Birthing Position. *American Journal of Public Health* 77 (5): 636–641.

Fildes, Valerie A. 1986. *Breasts, Bottles and Babies: A History of Infant Feeding*. Edinburgh: Edinburgh University Press.

Goyal, Praveen Kumar, and Andrew N. Williams. 2010. 'To Illustrate and Increase Chyrurgerie': Ambroise Paré (1510–1590). *Journal of Pediatric Surgery* 45 (2): 108–114.

Green, Monica H. 2009. *Making Women's Medicine Masculine: The Rise of Male Authority in Pre-modern Gynaecology*. Oxford: Oxford University Press.

Guillemeau, Jacques. 1612. *Child-Birth, or the Happy Deliverie of Women*. London.

Smith, Helen, and Louise Wilson (eds.). 2011. Introduction. In *Renaissance Paratexts*. Cambridge: Cambridge University Press.

CHAPTER 4

Graphic Surgical Practice in the Handbills of Seventeenth-Century London Irregulars

Roberta Mullini

INTRODUCTION: MEDICAL HANDBILLS AND THEIR CONTEXT

In the second half of the seventeenth century, the London medical marketplace saw a vast increase in the publication of advertisements highlighting a practitioner's surgical skills and selling products for healthcare. These adverts, distributed free of charge along London streets or in market squares, and also handed out at the feet of mountebanks' stages,[1] were printed for irregular practitioners. These men and women were not approved by the Royal College of Physicians because they lacked a university education and the required Latin literacy.[2] In order to print their adverts, practitioners relied on "small workshops, or corner presses" (Furdell 2002: 143), whose owners are rarely mentioned. The paper on which handbills were printed was flimsy and cheap, so that verso words often bleed through on the recto, and vice versa. The printing process, furthermore, was inaccurate, with poor or non-existent proofreading, evidenced by the abundance of misspellings in the texts.

R. Mullini (✉)
Università di Urbino Carlo Bo, Urbino, Italy

© The Author(s) 2018
H. C. Tweed, D. G. Scott (eds.), *Medical Paratexts from Medieval to Modern*, Palgrave Studies in Literature, Science and Medicine,
https://doi.org/10.1007/978-3-319-73426-2_4

57

While most handbills consisted of words tightly packed on two pages of a single sheet of paper, others showed a certain care for visual details intended to catch prospective readers' attention. These details might include the addition of images, various font types, decorative borders, manicules, portraits, coats of arms, alchemical instruments, and symbols. This chapter considers how such details interacted with the language of advertising (see Leech 1966: 28–29) in the early stages of mass communication and marketing procedures. Furthermore, this chapter examines what the interaction between linguistic and visual features can reveal about the producers and readers of seventeenth-century handbills.

When Daniel Defoe, in *A Journal of the Plague Year*, commented upon the typographic strategies of "ignorant Fellows; quacking and tampering in Physick" (1722: 36), he applied an innovative approach to the reading and understanding of the flyers distributed by irregular practitioners during the 1665 Great Plague. The multiple agencies behind Defoe's novel— the author, the unnamed printer, and the four publishers and booksellers—succeeded in mimicking the visual power and cogency of quacks' deceiving handbills by capitalising whole adjectives. So, a drink was "INCOMPARABLE", a remedy for the plague was "UNIVERSAL" and certain pills were "INFALLIBLE". Authors understood how the use of fonts of various sizes could influence readers' reactions and how powerful these techniques were in validating and legitimising the messages contained within the bills.

Both physicians and surgeons used visual features in their handbills, although surgeons were often more specific in their use of graphical elements when advertising their professional practice and skills. Some handbills advertised their writer's practice by reproducing images of operations on the page. The medicaments offered by irregulars were not so different from those that university physicians could prescribe, given that scientific medicine still relied almost exclusively on herbal remedies. Some handbills only extolled the individual author's practice and remedies, whereas others also attacked deceivers—rivals in the same medical domain, be they university physicians or irregulars (competition to gain clients could be fierce).[3] Although handbills were still used at the beginning of the eighteenth century, their role in marketing medicine decreased in the following decades, substituted mostly by advertisements in newspapers and periodical publications, after the latter's successful advent and spread (as discussed by Laura Mainwaring, in her chapter, "Profit and Paratexts; the Economics of Pharmaceutical Packaging in the Long Nineteenth Century", in this collection).

The Case Studies

The following section examines a selection of handbills held by the British Library in two dedicated collections.[4] An interdisciplinary approach is applied, including linguistics (pragmatics and the language of advertising), concepts derived from the study of popular culture and the social history of medicine, and from multimodal analysis.[5] The focus of this investigation is three advertisements, printed for three self-proclaiming doctors, all of whom practised in London in the last quarter of the seventeenth century. A point of special interest is the relationship between text and image; these handbills were addressed to both educated and illiterate people, a diverse readership sharing a common desire for remedies and medicaments.

The handbill issued by "J. Russel[l], Professor of Physick and Oculist" is the first to be encountered in the BL collection, shelf-marked C112f9 (Fig. 4.1). It is printed on both pages, and includes 2417 words, much above the average of 653 per bill.[6] Only 547 words are on the recto, making the verso densely packed with text and difficult to read. However, the scale and position of the images attract the reader's attention despite the letter-crammed verso.

Handbill C112f9[77] counts only 1403 words, distributed on the bottom half of the recto (448 words) and on the whole verso, while the top half of the recto is occupied by a single woodcut (Fig. 4.2). The practitioner advertising his remedies here is not named, but called "an Experienc'd, most Famous, German, Turkish, and Imperial Physitian" (recto) and "Imperial Operator and Doctor" at the very end of the text (verso).

Handbill 551a32[112] also has a single image on the recto, occupying about one-third of the page (Fig. 4.3). The text, which contains the name of the practitioner (Cornelius à Tilbourn, or Tilburg), consists of 538 words, of which 173 are on the recto. It is significantly shorter than the previous two, but it is considered together with them because here the interest does not lie in quantitative data (which are given in order to show how different handbills could be), but in their graphic arrangement and textual structure.

The historical context in which the three flyers were printed is very similar. Despite the impossibility of dating them precisely, they all belong approximately to the last two decades of the seventeenth century. All were distributed in London and their purpose was similar—to sell medicines, extol their authors' healthcare skills, and convince prospective customers. In such a setting, these handbills are clearly "semiotic resources" (van

Fig. 4.1   John Russell's Handbill (C112f9[1], recto) (By courtesy of the British Library/Bridgeman Images)

Fig. 4.2 Handbill C112F9[77], recto, detail (By courtesy of the British Library/Bridgeman Images)

Fig. 4.3 Cornelius à Tilbourn's Handbill (551a32[112], recto), detail (By courtesy of the British Library/Bridgeman Images)

Leeuwen 2005: 1–6), worth investigation as signs used to communicate the various contexts of everyday life.

THE HANDBILLS: TEXT AND VISUAL PARATEXT

"J. Russel[l], Professor of Physick and Oculist"

Given the visual impact of the unique recto page, it is evident that whoever designed the structure of Russell's handbill (C112f9[1]) had a clear marketing agenda.[7] The page is divided into two parts, one inside the other. The external section has twelve images of surgical operations plus the top centre portrait (of John Russell) and some objects connected to distillation. This structure builds a frame around the text, but it also integrates words, distributed as captions for each picture (excluding the practitioner's portrait). Lines are drawn to segment the page, disconnecting each image from the rest of the page, but at the same time also allowing links with one another through "rhyme" (for example, the same drawing strokes in the woodcuts).[8] Spaces are mainly rectangular (the two square ones at the bottom are parallel and symmetrical to the top centre images). However, one of the spaces is distinct; the centre top portrait has an oval frame inside a rectangular one, which highlights the image out and suggests that the elegantly periwigged person portrayed in his full dignity is John Russell.

The man is also singled out by other signifiers; there is a furnace on his right-hand side and some shelves with vases and alembics are on his left. These objects are instruments used in the distillation and conservation of substances for healthcare, thus connecting the materials of the profession to the "professor". The design may also be capitalising on the public's fascination with instruments associated with alchemy.[9] John Russell was evidently adept at self-promotion; another handbill of his exists with the same images but a different text (C112f9[148]), and another still (supposedly a later version) in which he states that "I should have forborn Publishing Papers if I had not been Reported Dead."[10]

All images display both male and female body parts and also full, naked bodies that are being operated upon. In spite of the title words introducing Russell as "Oculist", only two operations show ophthalmic procedures (the removal of a cataract from a man, and of an eye "carnosity" from a woman). The remaining images portray surgical treatments on other body parts. The images appear to depict five women, six men and (perhaps) a child. In this printed 'operating theatre', there is no prudery in exhibiting suffering and private parts. A woman's breast is being removed in the top right image, another's breasts and belly are exposed down on the right-hand side, and a male body is undergoing the repair of anal fistula. Naked male bodies feature in a diagonally symmetrical position: bottom right, a man is being operated on for a rupture (hernia), and, top left, another male is portrayed wearing a truss, that is, he has already been cured of hernia. There is no attempt to hide the suffering caused by the operations; in these illustrations of surgery there are blood drops coming out of the cuts, and tears from some patients' eyes.

The captions in each vignette name the individual operations: "A Carnosity Cutting off the Eye"; "Couching a Catract [*sic*] of one that had been blind 30 years"; "Fistula in the Fundament"; "A Rupture Cured". Text and image, therefore, are integrated (van Leeuwen 2005: 13), mirroring each other visually and verbally. Some questions arise from the overall arrangement of the page: why are these images so graphic?; and why are there captions sharing the image space? The answer to the first question involves the issue of the images' truth value. Late seventeenth-century readers of course knew that surgeries were painful and that blood was spilled during the procedures; nevertheless, these realistic representations of surgical operations might have increased the truth value of the handbill as a whole. The image is encountered first and the text is analysed later, thus accruing the latter's perlocutionary force (see Austin 1962). A possible answer to the second question might be that images function to those already in the know as a cajoling and flattering tool, or—to the less literate—as the correct naming of something the addressee knows but cannot specify. On the one hand, these captions might be interpreted as a sign of the author's power and knowledge in the medical domain and, on the other, as a blandishment of the reader's education (as if the author were addressing what linguistics calls 'positive face', and saying to the reader: "I confirm your existing scientific knowledge").[11]

The narrative at the centre of the frame is divided into indented paragraphs which show a certain care and writing knowledge on the part of the author and printer. Larger fonts are used for the practitioner's name and specialisation, smaller ones for his address in the conspicuous incipit position, and the smallest for the narrative. At the bottom of the recto page, contrasting with the bulk of the writing, there is the phrase "I Draw Teeth with a Touch" printed in the same font size as Russell's address: the blank space separating it from the rest singles it out and reinforces the narrative, inviting prospective patients to rely on this "Professor of Physick" even for such a 'simple' operation as a tooth extraction.

Roman type is used especially on the recto, where italics are limited to the author's name and address, and to the Latin names of some anatomical parts of the eye (e.g. "*Aqueeus Vitreous* or *Albugineous* Humours"). Otherwise on the verso italics are much more present and also used for the intertitles fragmenting the text and for a whole short paragraph in which the author assures patients suffering from the French pox (syphilis) of the beneficial effects of his cure. These words are highlighted by their font because of their explicit and implicit meaning: the sentence "*I seldom hinder any Business, and cure so privately*" implies Russell's discretion in hiding the consequences of promiscuous sexual behavior. The italicised paragraph comes at the end of the text section entitled "*Symptomes of the* POX" (verso) which is accompanied by three other titled paragraphs, each of which extols specific products.[12] The entire verso is divided into two columns down to one-eighth of the page, where a horizontal line segregates what precedes from what follows—that is, the list of London retailers of Russell's medicines—an effective use of space and layout.

The handbill as a semiotic resource—besides its fundamental marketing purpose—shows the priority conferred to the visual mode over the verbal one. This is clearly visible from the images' disconnectedness from parts of the verbal content. In certain cases the primacy of the visual elements is conveyed by their appearance in spite of the absence of references in the content of the narrative—for example, nowhere does the text speak of ruptures or of wens, or of any operation in the skull, whereas some images depict these surgeries. While words are generally in small-sized fonts, images are quite prominent, and in the case of John Russell ,the "Professor of Physics", the message seems clear—he relies more on his surgical skills than on his medical expertise.

The "Experienc'd, Most Famous, German, Turkish, and Imperial Physitian"

The image in C112f9[77], framed by a black thick line, occupies half of the front page. It includes some objects, four human figures and the words "This High-German Doctor, cured the Emperor of the Turk's Brother, who was Blind 13 Years". The verbal mode accompanies the visual one

with which it is integrated, at least in this part of the bill. The words cannot simply be defined as a caption, since the verb is in the past, thus
acquiring rather the flavour of a narrative than of a description. In any
case, this short narrative, foregrounded by its large font size, offers cues to
the understanding of the protagonists of the image and, by doing so,
records the most important event of which the author boasts, an event
also reworded in the longer narrative below. At the very top of the page
there are some words that anticipate the image; printed with even larger
fonts the reader finds the phrase "With Liberty of the Colledge of Phisitians
of the Royal Head City of *LONDON* in *ENGLAND*" in two lines. It suggests the author's status is higher than other practitioners', given the "liberty" obtained from the Physicians' College. Nothing is known about the
practitioner's name, meaning even a search in the College Annals could
not prove the claims true or false, and we, albeit conscious of the high
dose of possible overstatement of the phrase itself, must warily accept
what the writer says.

The vignette shows four long-moustached men, two sitting on arm-
chairs and two standing, in a curtained room adorned with a tiled floor.
On the right-hand side one of the men is holding a sceptre in his right
hand and has a crown on his turbaned head—he is the "Emperor of the
Turk" mentioned in the words inside the woodcut. Behind him on a low
table lies what appears to be a sword, plus a circular object from which
originates an unidentifiable long and coiling piece of cloth, perhaps the
patient's turban taken off just before the operation. The other personages
are gathered in the left part of the image: a standing turbaned man is push-
ing one of the sitters' forehead backwards, with both hands, while the
other sitter, near the left border of the image, is evidently the surgeon,
looking attentively at his patient and holding a lancet in his right hand,
ready to operate. The patient, the 'emperor', and the other standing man
wear identifiably Turkish garments. The top border is porous to let the
emperor's crown be clearly identifiable.

Here, as well as in John Russell's handbill, the representation appears to
play a high social semiotic role; to late seventeenth-century readers the
image testifies to the truth of what the operator writes, augmenting the
appeal of his credentials and the reality value of his statements in the rest
of the advertisement. This is despite the exaggerated first paragraph of the
following narrative, where this "High-German Doctor" claims to be:

an Experienc'd, most Famous, *German, Turkish,* and *Imperial* Physitian, whose like has not been in this Kingdom, who hath Learned such a curious and strange Art, which no other Doctor doth understand, and can Cure all sorts of Patients which are left off by other Doctors. (recto)

From this quote one can notice at once how italics are deployed for emphasis. The first paragraph of the text is followed by numbered paragraphs, the first two on the recto and the remainder on the verso—that is, the narrative is segmented fairly regularly, each numbered paragraph focusing on specific illnesses. In the first paragraph the writer does explain his ability in ophthalmic surgeries, in this way linking words to the image, but thence nearly all diseases known at the time are mentioned, from "ruptures" (par. 2), to "Hair-Lips" [*sic*] (par. 3, verso), to the "French Pox" (par. 4), to female barrenness (par. 5), to deafness (par. 6), to the "Falling Sickness" (par. 7), up to a list of sixteen other ailments (par. 9). Not satisfied with what he has listed so far, the writer also adds "This Famous Physitian knoweth to cure all other distempers which are not mentioned here" (verso). Rather than mentioning medicines, this practitioner insists on his "curious art" (the phrase occurs three times) and on his ability which goes "beyond other Doctors"; in other words, he presents himself as the best physician and surgeon in town, although without any specialisation.

How aware this self-proclaimed doctor was of the convincing power of print is shown by the final paragraph:

This Imperial Operator and Doctor is at present in LONDON, and now liveth at a New House the corner of White-Cross-Alley in Moorfields, next door to the Star Musick-House, where you may see the High-German Spread Eagle hang over the Door. (verso)

The operator's address is printed in black letter, interspersed with some words in roman type. While John Russell exploits the initial position for such a relevant piece of information, this 'doctor' takes advantage of the final lines of his handbill, both incipit and explicit being places which certainly attract readers' eyes sooner than any other part of a text. Furthermore, the use of black letter—and of the blank space separating the quoted paragraph from what precedes it—highlights the semiotic role of the paragraph itself; printer and author count on the traditional associations of this font, typically employed in ballad printing, for royal documents and for all printed matter 'of old'.[13] Black letter is used to grant authority to this practitioner and, at this point, readers knew where they had to go if they wanted to be cured.

"Cornelius à Tilbourn, the Famous German Physitian and Operator"

Cornelius à Tilbourn's handbill 551a32[112], one of his eight adverts still extant in the two BL collections, shows yet another distribution of the page space. The recto can be divided into three parts; the top (more or less a third of the page) is occupied by an image, the centre by an indented paragraph using italic and roman fonts, the bottom third by another paragraph starting the narrative proper, in smaller font.[14] Space is thus well separated in three blocks; there is no integration of text and visual part, and the two verbal sections are set apart and framed by a blank.

This image also represents an ophthalmic surgery taking place in an aristocratic inner space. A curtain hangs on the right-hand side, a paned window is visible at the back, there is a carpet on the floor and a low table or a cupboard on the left. Through an open door a child is looking at what is going on. Centre stage, a man (or maybe a woman if what protrudes from the sitter's head is a female periwig) sitting in an armchair is being operated on by a long-wigged and authoritative-looking man on the patient's right.[15] The operator has an unidentifiable instrument in his left hand. The function of this image and the explanation have to be found in the text. Readers cannot immediately ascertain the relationship between text and image; they must read beyond the large bold capitals and the initial words, where Cornelius states that, having "lately found out some admirable Remedies, which was never yet made Publick in any of his former Bills":

> These are therefore to Advertise all Persons concern'd, that have either Weak or Dimm Sight, occasion'd by Age or otherwise, and are oblig'd to wear *Spectacles*, He [Cornelius] undertakes to bring them to see well without them in a Weeks time, altho' they have used them 20 Years before: He Infallibly Cures all sorts of Sore Eyes whatsoever ... (551a32[112] recto)

The elegance of the woodcut seems to serve exclusively the content of the narrative, i.e. it does not play the role of visual summary of a specific event (as in the previous handbill), or as an exhibition of the "manual operations" John Russell is able to perform (C112f9[1] recto). It can be read as an affirmation of Cornelius's identity as an oculist. The relatively short text actually deals with ophthalmic problems in its first two paragraphs (223 words out of the total 538), but the operator also adds other ailments he claims to be able to cure (miscarriage, venereal diseases, "fainting fits"), thus presenting himself as a generalist rather than a specialist, in contrast to the message of the visual part of the advertisement.

But there is something in the text which differentiates it from the two others. Cornelius à Tilburg proclaims the official titles he is proud of—he is "Sworn Chyrurgeon to the late King Charles the II, And now Priviledg'd by Our Gracious Soveraign King William" (recto). His irregularity as a practitioner is redeemed by his office as "Sworn Chyrurgeon" to a king—the "Emperor of the Turk" introduced in handbill C112f9[77] as an exotic testimonial is replaced here by two English sovereigns, nearer the readers, and more 'verifiable'. The social meaning of this change counts on the values of national identity and proximity, on something readers might have been able to check, instead of relying on the exotic and the foreign.

CONCLUSIONS

The three cases presented here show that irregular practitioners understood the marketing potential of illustrated handbills. Surgeons in particular took advantage of the exhibition of their 'operating theatres', focusing on the specialisation they considered most appealing to prospective patients. In the cases studied, the adverts are addressed to a primarily urban readership and ophthalmic surgeries appear frequently in irregular practitioners' handbills.[16] It is necessary, though, to distinguish between the striking but simple visuals in the "Imperial Physitian"'s handbill and in Cornelius à Tilbourn's, on the one hand, and that of John Russell, on the other. One may wonder whether the abundance of images in Russell's advertisement and especially their anatomic details stem from this practitioner's personal competence, or from his skill in availing himself of a repertoire of surgical representations.

Russell's figures are not original; "professor" Russell took most of his images from *Armamentarium Chirurgicum*, a text originally printed in Latin at Ulm (Germany) in 1655 by Dr. Johannes Scultetus and translated into English in 1674 with the title *The Chyrurgeons Store-House*. By comparing Russell's images to Scultetus', at least nine images from the illustrated frame appear to derive straight from the German doctor's work.

It is impossible to know whether Russell read the original publication or its English translation. Certainly he had some of the finely engraved images from Scultetus' tables reproduced roughly. Besides not mentioning the source of the visual apparatus, Russell produced an edited version of many figures. For example, the woman whose breast is being cut is not weeping in the original (Fig. 4.4); Russell added this detail. The operation of the "fistula in the fundament" makes use of a picture from Scultetus' Table XXXXI (1674, fig. vi, 194) which the original author describes quite differently as representing the perforation of the "ars gut of children newly born being shut" (197).

The cutting of the enormous wen in Russell's bill corresponds to image vi in Scultetus' Table XXXIV (140), but its meaning is quite different. No wen is removed in Russell's bill, but an operation is performed (142). Russell copied the image and enlarged the tumour, rendering it more

Fig. 4.4 Johannes Scultetus, *The Chyrurgeons store-house*. Table XXXVIII, "Amputation of the breast", detail (image n. L0005307, from the 1672 Amsterdam edition of *Armamentarium Chirurgicum*) (By courtesy of the Wellcome Library, London)

conspicuous. He was guilty of plagiarism according to our contemporary standards, but he was also able to effectively exploit graphics and print technology.

The writers of the three handbills investigated in this article understood the communicative power of images and used them as business cards to be handed out to the general public. In the text they explained their many-faceted medical profession, but to the visual mode they entrusted the success of their surgical practice. There is a fundamental cohesion between the modes used in these advertisements. However, it is not easy to read the images as paratextual features and the wording as 'the' text; the latter was likely to be read only because of the former, which would acquire then the status of main text. Words, in this way, would become a sort of appendix to visuals and the usual relationship between the two is reversed.

NOTES

1. Mountebanks sold quack medicines from platforms where music and other entertainments were also performed. The case of John Wilmot, second earl of Rochester, is famous: around 1676, after being banished from court, he mounted a stage in the City and, disguised as a mountebank under the pseudonym of Dr. Alexander Bendo, performed 'cures' for some months (see Mullini 2015: 197–203).

2. Irregular practitioners were not necessarily illiterate: some of them had university education, although the College of Physicians seldom acknowledged medical degrees when obtained abroad, even if in famous continental medical schools.

3. Irregular practitioners have been studied, among others, by Thompson (1928), Porter (1989), and Pelling (1998, 2003).

4. See the Bibliography. Individual adverts will be quoted by their collection shelfmark, followed—between square brackets—by the number pencilled on each of them.

5. For brevity I refer to the seminal works of Austin (1962) and Brown and Levinson (1987) for pragmatics; to van Leeuwen (2005) and to Kress and van Leeuwen (2006) for multimodality.

6. This results from the special corpus I created for a previous research by transcribing 307 of the 416 leaflets: the total amount of words is 200,583 (see Mullini 2015).

7. In the British Library collections there is only one handbill whose structure resembles Dr. Russell's: it is C112f9[126], selling "The Famous Water of Talk and Pearl", but no surgical operation is shown there.

8. van Leeuwen (2005: 13) speaks of "rhyme" when "two elements, although separate, have a quality in common ... a colour, a feature of form ... etc.".

9. Distillation tools were used by apothecaries to prepare herbal remedies, but also by those who, following ancient philosophical and cabalistic thinking and occult traditions, tried to turn cheap metals, such as lead, into gold. Alchemy, though, was also the basis of chemistry which, especially after Paracelsus (1493–1541), started to produce chemical remedies for health-care, even though magic and esoteric belief remained connected to it.

10. *J. Russel, professor of physick, and oculist*, Bodleian Library. The same frame (but after the substitution of two images) was used by a later practitioner (Harley 5931[76]): this reveals, at least, that this page layout was so successful that it was employed as a brand.

11. The positive face is defined as a person's desire to be appreciated by others. For this concept, see Brown and Levinson (1987: 61).

12. Very probably because of the relevance of venereal diseases and of the pressing demand for cures for them in late seventeenth century (see Siena 2004), this paragraph is signalled by a manicule, thus attracting the reader's eye sooner than other parts of the narrative.

13. Around 1590, roman type superseded black letter (see King 2013), but in the last quarter of the seventeenth century the title-pages of some medical treatises included at least one black-letter line (see images in Taavitsainen and Pahta 2010, CD ROM, "EMEMT Gallery"). On the semiotic and social meaning of black-letter texts in the period as "typographic nostalgia", see Lesser (2006: esp. 107).

14. On this practitioner, see Thompson (1928: 86–90), Matthews (1964: esp. 36–39), and Mullini (2016).

15. It is well-known that wig-wearing became fashionable during the Restoration as a status symbol.

16. In the corpus (see note 6), the words "eye" and "eyes" (connected to the cures offered by medical practitioners) occur 49 and 142 times, respectively. "Eyes" ranks fifth in a 25-item list of the body parts most frequently mentioned in the adverts, after "head", "stomach", "reins", and "teeth", which are present 473, 223, 212, and 167 times respectively. The list was obtained by processing the corpus with a concordancer.

Bibliography

551a32: *A Collection of 231 Advertisements*, etc. *The Greater Part English and Chiefly Relating to Quack Medicines, the Rest German Descriptions of Commemorative Coins and Medals* [1675-1715] (BL shelfmark).

Austin, John L. 1962. *How to Do Things with Words*. Oxford: Clarendon Press.

Brown, Penelope, and Steven Levinson. 1987. *Politeness. Some Universals in Language Usage*. Cambridge: Cambridge University Press.

C112f9: *A Collection of 185 Advertisements, Chiefly Relating to Quack Medicines. The Greater Part English, the Rest French, German and Italian* [1660-1716] (BL shelfmark).

Defoe, Daniel. 1722. *A Journal of the Plague Year*. London: Printed for E. Nutt; J. Roberts; A. Dodd; and J. Graves. Accessed via Eighteenth Century Collections Online.

Furdell, Elizabeth Lane. 2002. *Publishing and Medicine in Early Modern London*. Rochester: University of Rochester Press.

Harley 5931. J. Bagford, *Collection for the History of Printing* (BL shelfmark).

King, John N. 2013. Introduction. In *Tudor Books and Readers: Materiality and the Construction of Meaning*, 1–14. Cambridge: Cambridge University Press.

Kress, Gunther, and Theo van Leeuwen. 2006. *Reading Images. The Grammar of Visual Design*. Milton Park: Routledge.

Leech, Geoffrey N. 1966. *A Linguistic Study of Advertising in Great Britain*. London: Longmans.

Lesser, Zachary. 2006. Typographic Nostalgia: Play-Reading, Popularity, and the Meanings of Black Letter. In *The Book of the Play: Playwrights, Stationers, and Readers in Early Modern England*, ed. Marta Straznicky, 99–126. Amherst/Boston: University of Massachusetts Press.

Matthews, Leslie G. 1964. Licensed Mountebanks in Britain. *Journal of the History of Medicine and Allied Sciences* 19 (1): 30–45.

Mullini, Roberta. 2015. *Healing Words. The Printed Handbills of Early Modern London Quacks*. Frankfurt: Peter Lang.

———. 2016. 'I Cornelius à Tilbourn': Hotchpotches, Poisons, Antidotes and Royal Gifts. In *Chlorophyll Killers*, ed. Jan Marten I. Klaver and Giuseppe Puntarello, 119–148. Fano: Aras.

Pelling, Margaret. 1998. *The Common Lot: Sickness, Medical Occupations and the Urban Poor in Early Modern England*. London: Longman.

———. 2003. *Medical Conflicts in Early Modern London: Patronage, Physicians and Irregular Practitioners 1550–1640*. Oxford: Clarendon Press.

Porter, Roy. 1989. *Health for Sale: Quackery in England 1650–1850*. Manchester: Manchester University Press.

Scultetus, Johannes. 1674. *The Chyrurgeons Store-House: Furnished with Forty Three Tables Cut in Brass, in Which Are All Sorts of Instruments, Both Ancient and Modern; Useful to the Performance of All Manual Operations*. London: Printed for John Starkey. Accessed via Early English Books Online.

Siena, Kevin P. 2004. *Venereal Diseases, Hospitals and the Urban Poor: London's "Foul Wards", 1600–1800*. Rochester: University of Rochester Press.

Taavitsainen, Irma, and Päivi Pahta, eds. 2010. *Early Modern English Medical Texts*. Amsterdam: John Benjamins.

Thompson, C.J.S. 1928. *Quacks of Old London*. London: Brentano's.

Van Leeuwen, Theo. 2005. *Introducing Social Semiotics*. Milton Park: Routledge.

CHAPTER 5

Profit and Paratexts: The Economics of Pharmaceutical Packaging in the Long Nineteenth Century

Laura Mainwaring

This chapter uses packaging as a means of analysing the role of paratext in shaping the economic environment of the medical marketplace in the long nineteenth century. It outlines how developments in commercial legislation within a wider economic context shaped the paratext found on these packages. In order to highlight the significance of packaging as a medium for paratexts, I undertake the first quantitative survey of the visual depiction of packaging within advertisements in the influential nineteenth-century trade periodical, *The Chemist and Druggist*. As a trade periodical, the target audience of *The Chemist and Druggist* was retail and manufacturing chemists. The retail chemist acted as an intermediary "consumer" between the manufacturer and the end-consumer, and their importance should not be overlooked. Their significance is evident from surveying the retail space—even though stock was on display this did not mean that the public were allowed to self-select goods or products. A retired chemist, Alan Garrett, an apprentice in the 1920s, commented that early on in his career:

L. Mainwaring (✉)
University of Leicester, Leicester, UK

© The Author(s) 2018
H. C. Tweed, D. G. Scott (eds.), *Medical Paratexts from Medieval to Modern*, Palgrave Studies in Literature, Science and Medicine,
https://doi.org/10.1007/978-3-319-73426-2_5

Everything was behind glass doors, not help yourself, you had to open the cupboards to get out what you wanted, and customers didn't help themselves. Even the counter was closed, no things to pick up off the counter, there was a shelf along the top which some things were displayed on, but there was no self-selection of any kind at all.[1] (Anderson 1995)

The retail chemist thus had an element of control over the consumer's purchase, which suggests that advertisements aimed at the trade should have greater significance within the history of medicine (Jones 2013). Hilary Marland (2006: 87) has argued that, for most chemists and druggists, direct trade with the public was more significant than the network of trade with the medical profession. Indeed, a wholesaler or manufacturer might persuade the retailer to display and stock a certain product via the draws of a particular brand or aesthetic package, which was then sold onto the lay consumer (Church 2000: 627).[2]

Packaging provides the physical representation of the contents inside. A typical package for a healthcare product from the long nineteenth century would commonly be made up of separate components with marks from various market participants: a glass bottle embossed with the manufacturer's name; a paper label with the logo, trade mark, or coat of arms of the proprietor; a medicine stamp from the stamp office;[3] the name and address of the dispensing chemist; and perhaps a poison label from the retailer to indicate any dangerous contents in line with institutional regulation.[4] A court case that took place in the Old Bailey will be discussed later on in the chapter, as an example of how paratext and branding on packaging were formed via a chain of participants and disseminated in the medical marketplace. Paratext in this chapter, defined as a form of graphic communication and/or supplementary text, includes the container itself, labels, surface markings, stamps, seals, wrappers, pamphlets, and the colours used. These paratexts could transmit information about the commodities' origins as well as an understanding of the contents, such as its dosage and risk. These constituent parts constructed the identity of the product, acting as forms of branding.[5]

Commercial Legislation and the Changing Nature of Paratexts

The history of branding is usually associated with the factory age and mass-mechanised goods, but branding can be traced back to earlier times. Patent medicines were one of the first commodities to be distributed on a national scale (Doherty 1992; Styles 2000). From the seventeenth century, the use of trade marks was used for the promotion of many of these remedies, and any judicial proceedings were carried out under common law (Bently 2014). However, the nature of the trade mark was altered with the passing of the Trade Marks Registration Act of 1875, with the creation of a register for proprietary marks for the first time (Bently 2008). Under this Act, the trade mark was defined as a visual mark, akin to a logo, "printed in some particular manner" or written distinctly (Price and Swift 1988: 3). From 1876, these legally protected marks started to appear in advertisements and on packaging, in the form of an illustrative logo or typographic text; this added a visual reference to aid brand recognition and enabled the proprietor to reinforce their business image through graphic communication. *The Trade Marks Journal* advertised the mark intended for registration, allowing for objections to be issued and marks to be withdrawn. This process was deemed to hold significant authority for vendors of medicines. For example, Arthur Cox H and Co, Tasteless Pill Manufacturers, remarked in an advertisement (*The Medical Press & Circular* 1880: 10) that:

> The registrar of Trade Marks has just granted us, after three months' publicity, the Trade Mark of which the above is a facsimile, thus officially recognising our claim the "ORIGINAL MAKERS OF TASTELESS PILLS." All packages sent out from this date will bear this mark.

Price and Swift (1988) collated a catalogue of medically related trade marks from the *Trade Marks Journal* from its inception in 1876 up until 1880, listing a total of 1160 products sold for healthcare-related needs.[6] Their catalogue shows the range of styles used for this paratext; an array of typographical fonts, elaborate illustrations, labels, as well as signatures that reflect the simple form of branding found in the early medical marketplace. A number of trade marks for healthcare products featured a container as part of the logo, and some logos were made up entirely of a template of the product's packaging.

The industrial age caused major economic shifts by separating the producer and the end-consumer in the distribution chain. Consumers could no longer rely on personal exchanges with the manufacturer to assess the reputation of a vendor. This was a particular problem in the medical marketplace; healthcare products were high-risk commodities within a market made up of poison scandals and ineffective therapeutics (Whorton 2011; Haller Jr. 1975). Mira Wilkins (1992: 20) has explored how the trade mark evolved into a valuable asset in the nineteenth century through its ability to transfer across the intangible qualities of reputation and identity, increasingly important within a depersonalised market. The impact of the Act and its successive revisions can be gauged from advertisements found within medical trade catalogues and periodicals, with a proliferation of advertisements from 1876 explicitly announcing products' new registered status, like the one shown in Fig. 5.1 (*The Chemist and Druggist* 1890: 15). Many of the products noted as registered were actually only registered through the Stationer's Hall (see Fig. 5.5), which did not give them a protected trade mark status. *The Chemist and Druggist* ("Notes and Queries" 1875: 250) remarked to their readers that "Some people have a fancy for registering the title of their preparations at Stationers' Hall, which costs them 5s. Id. We cannot say what direct benefit results from that expenditure." They also exclaimed that many of their readers think "there is some legal virtue in registration at Stationers' Hall," but "that will not be taken as evidence" if someone else registers the same mark as a trade mark in that class of good ("Notes and Queries" 1875: 436).

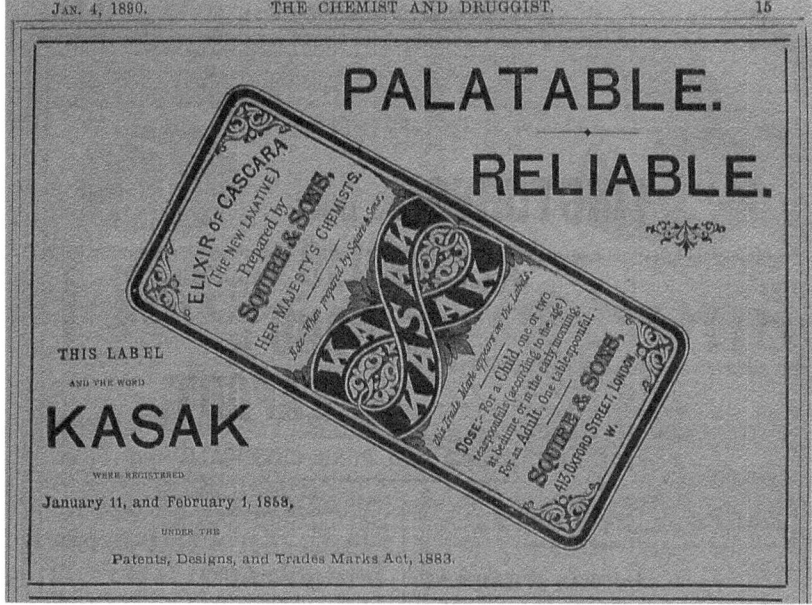

Fig. 5.1 Advertisement of Squire & Sons, with "Registered Trade Mark" as a paratext on the label of a laxative, 1890 (Wellcome Library, London)

Proprietors can thus be seen to be using the "registered" paratext as a means to enhance credibility within the market, regardless of whether the mark had been registered as a trade mark.

Packaging as an important medium for the promotion of a healthcare product's "registered" trade mark and other layers of branding became more pronounced during the latter quarter of the nineteenth century; advertisements from companies selling packages, boxes, and printing services to the retail and wholesale trades were a common occurrence in *The Chemist and Druggist* during this period.[7] Depictions of a product's packaging or label within advertisements rose steadily from 1880 to 1905. A quantitative analysis of the advertisements found in *The Chemist and Druggist*, taken from a sample of 2555 advertisements from 11 issues between 1880 and 1905, demonstrates an increase of the graphic display of packaging and labels for the promotion of healthcare products (see Fig. 5.2).[8] Advertisements comprising images of packaging rose steadily from just 4% of advertisements after the introduction of the Trade Marks Act in 1875, with two peaks in the wake of the 1887 Merchandise Marks Act, and rose up to 21% by 1905. This last peak coincides with the passing of the 1905 Patent Designs Act.[9]

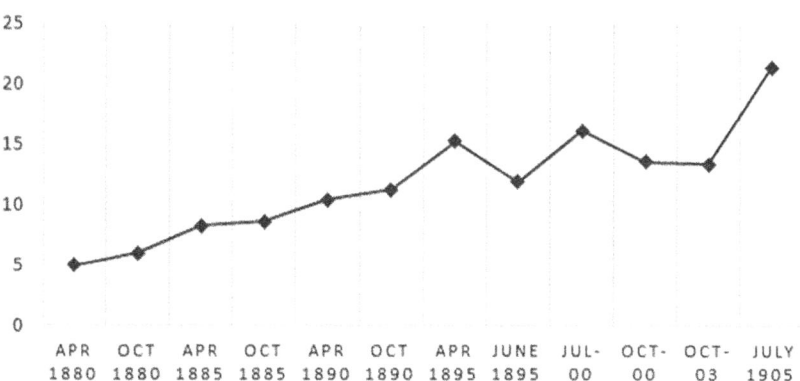

% OF ADVERTISEMENTS DEPICTING PACKAGING OR LABELS

Fig. 5.2 Graph showing the rising number of graphic illustrations of packaging and labels in advertisements in *The Chemist and Druggist* from selected issues between 1880 and 1905

The employment of packaging and labels in advertisements assisted the act of consumption by making the product and its paratexts more visible to the chain of consumers. Figure 5.2 suggests that packaging increased in significance as a medium for transmitting branding during the last quarter of the nineteenth century when Trade Mark Law was formed.

Paratext in Product Advertisement

The use of graphic communication in advertisements allowed the physical appearance of standardised products to be employed as a means to assert credibility and genuineness. Text-only advertisements had previously used descriptions of packaging and labels as an assurance of genuineness prior to the proliferation of pictorial advertising, however, the use of packaging to engender trust could now be fully realised with illustrative advertising. For packaging to become a useful tool for consumers to decide whether they were buying the genuine article, the product and its paratext would have to look identical to the depictions found in the advertisement, otherwise it rendered the caution of imitation obsolete. Rendell's "Wife's Friend" Soluble Quinine Pessaries were advertised with a warning against imitation, "To distinguish the genuine Pessaries from these fraudulent concoctions, see that the Registered Trade Mark and Number, in RED INK, are upon each box."(*The Chemist and Druggist* 1900: 51). The graphic depiction of what was asserted in the advertising copy gave the paratext more weight in the assurance of credibility (see Fig. 5.3).

Figure 5.4, an advertisement for Henry's Calcined Magnesia from *The Chemist and Druggist* (1910: xxiii) reflects the complexity of branding and packaging in the long nineteenth century medical marketplace. It reveals that the wrapping as well as the physical vessel were important ways to brand medicines, particularly with the ability to use them as pictorial devices within advertising copy. This advertisement highlights the layers of branding that a product could achieve via multiple paratexts when at risk from imitation: from the embossed bottle, the pasted label, the paper wrapper, the registered trade mark, the sealed bottle, to the Government stamp. Medicine stamps were in the form of a paper label and applied to a product to show that the vendor had paid the correct tax required for products with secret ingredients. Other than the required text, the medicine stamp could be blank, but it commonly featured the name and address of the proprietor; The Stamp Office could issue medicine stamps for individual proprietors with their mark shown on the label, which acted as a

JULY 7, 1900   THE CHEMIST AND DRUGGIST   51

RENDELL'S

" WIFE'S FRIEND " SOLUBLE QUININE PESSARIES.

THE WORLD-WIDE REPUTATION that these Pessaries have obtained since their invention by W. J. Rendell in 1885 has induced unscrupulous persons to place upon the market worthless and injurious compounds, with colourable imitations of W. J. Rendell's labels. To distinguish the genuine Pessaries from these fraudulent concoctions, see that the Registered Trade Mark and Number, in **RED INK,** are upon each box thus:—

| Registered Trade Mark 🖙 without which none are genuine. No. 182,688. | *W. J. Rendell* |

Registered also in Australia, India, The Cape, The Argentine, Germany, &c.

🖙 IMPORTANT NOTICE AND CAUTION. 🖙

Several Injunctions in the High Court of Justice having already been obtained against "Infringers" (reports of which have appeared in "THE CHEMIST AND DRUGGIST," the most recent cases being March 19, 1898, and January 19, 1899), W. J. RENDELL'S SOLICITORS are instructed to take proceedings against all persons (Makers or Vendors) fraudulently using the name of "W. J. RENDELL," "RENDELL," or any colourable imitation thereof, in connection with Pessaries NOT manufactured by

W. J. RENDELL,

Inventor and Sole Maker of the "WIFE'S FRIEND" SOLUBLE PESSARY,

15 CHADWELL STREET, CLERKENWELL, LONDON, E.C.

Fig. 5.3 Advertisement for Rendell's "Wife's Friend" Soluble Quinine Pessaries in *The Chemist and Druggist*, 1900 (Wellcome Library, London)

form of branding for the proprietor (Basford 2012: 143–144). The proprietors of Henry's Calcined Magnesia chose to depict exactly how the product would be seen in the shop, with the Government stamp taking centre stage:

> The Purchasers of this article are requested to observe that the words, "Thos. And Wm. Henry, Manchester", are engraved on the Government Stamp, pasted over the cork of each bottle. This is the only effectual security against the counterfeited imitations, which are similarly moulded, under their names—the Trade Mark—"Henry's Calcined Magnesia." (*The Chemist and Druggist*, 1910, xxiii)

Logos, coats of arms, trade marks, signatures, and provenance marks all communicated messages from the numerous participants involved in the manufacture, production and dissemination of the product to the

ESTABLISHED A.D. 1772.

HENRY'S
CALCINED MAGNESIA.

For **HEARTBURN, HEADACHE, GOUT, BILIOUSNESS, ACIDITY OF THE STOMACH,** TRY

Henry's Calcined Magnesia

FREE FROM TASTE, SMELL OR
ROUGHNESS TO THE PALATE.

ADULTS.
A Safe Aperient for the most delicate constitutions.

CHILDREN.
May be given to children in early infancy. The addition of a small quantity of the Magnesia prevents milk turning sour on the stomach.

MESSRS. THOMAS and WILLIAM HENRY, of Manchester, England, beg to inform the Trade generally that they continue to manufacture their old-established "CALCINED MAGNESIA" in the Greatest Chemical Purity, and also wish to warn Buyers against the numerous spurious and very inferior imitations offered in various foreign countries.

For the guidance of the Trade they submit facsimiles of their Bottle, wrapped and unwrapped.

THE LABELS CAN BE HAD IN ENGLISH, SPANISH, PORTUGUESE, OR ITALIAN

Messrs. HENRY will be pleased to hear from Merchants who are suspicious of any Stock in their possession.

THE MAGNESIA MAY BE HAD THROUGH ALL WHOLESALE HOUSES.

Registered Trade Mark:—"HENRY'S CALCINED MAGNESIA."

Messrs. THOMAS & WILLIAM HENRY, 11 East Street, St. Peter's, MANCHESTER.

Fig. 5.4 Advertisement for Henry's Calcined Magnesia, *The Chemist and Druggist*, 1910 (Wellcome Library, London)

each stage of the product's life-cycle, from retail chemist to the end consumer. Medicine stamps, on the other hand, communicated information straight from the state to the consumer. Basford (2012: 124) states that Government stamps complicate our understanding of branding, suggesting that they acted as a promotional tool for the state by enhancing its power and authority. These stamps were not actually regulatory marks, but were adopted by some vendors as having been sanctioned by the Government. *The Medical Times Gazette* (1875: 656) shared concerns over the stamp to readers, "The regulation of the sale of quack medicines is very necessary. At present it is possible to sell most any poisonous concoction as a patent medicine, apparently under the direct sanction of the Legislature, evidenced by the Government stamp." *The Times* (1884: 9) reported that "there can be no doubt that many patent medicines are purchased under the assumption that the Government label is a sort of Hall-mark guaranteeing the purity of the compounds sold under it ... But the label really guarantees nothing." This appropriation of a mark for something other than its original intention shows that the paratext of a package could hold multiple meanings, and, in this case, troubling consequences for the vending of a pharmaceutical product. In 1885, the words "No Government Guarantee" were added to the stamp to protect against these claims of credibility (Anderson 2005: 237). This required alteration of text stresses the power of paratexts in communicating extra information not explicitly stated in the text of the product.

The exact wording of text featured on pharmaceutical labels was of importance when taking into account the fiscal regulation of the preparation. If ailments were mentioned on the packaging or label, then the medicine stamp duty was payable, however, if the body part was referenced, the preparation was not liable. For example, headache powders were liable, but head powders were not (Matthews 1986: 2–5). *The Chemist and Druggist* ("Patent Medicine Stamps" 1885: 724) remarked to their readers that,

> We have received a large number of letters during the past month from all parts of the country testifying to some extra activity on the part of the Inland Revenue officers, and to a consequent widespread panic on the part of chemists in regard to the labelling of some of their specialties. We have been invited to give our opinion as to the liability or otherwise of scores of preparations the labels for which have been sent to us.

This correspondence informs us about the relationship between commercial health products and paratext, and implies that monetary concerns about paratext found on labels and packaging started a conversation about the ways these labels could be altered to avoid fiscal regulation. Figures 5.4 and 5.5 show the prominence of the medicine stamp for shaping the materiality of these preparations (Boots Price List 1911: 88).[10] This source, however, has usually been left out of discussions of medicine branding, particularly with standard textual analysis (Brown 1976; Cody 1999). Barker (2009) for example, identifies the stamp as part of a common marketing method, but does not identify it as a form of branding appropriated by the proprietor. Thus, it is argued here that the integration of non-textual source material into an analysis of branding allows for a broader understanding of healthcare products in relation to state and commercial legislation.

Fig. 5.5 The Inland Revenue Stamp added to the materiality of the product. It also acted as a security seal. Packaging for Goulding's Teething and Cooling Preparation from the price list of Boot's Cash Chemist Ltd., 1911 (The Boots Company plc)

Paratexts and the Multiplicity of Market-Processes

As noted in the previous sections, the depersonalisation of society, with the loss of strong community ties in an ever-increasing urban world, meant that the nature of establishing trust was changing from strong personal relationships to more numerous, looser ties (Barker 2009: 381). Various marks found on the product like the trade mark and the Government stamp could be used to establish credibility in lieu of direct contact between market participants.

A court case that took place on 5 April 1886 at the Old Bailey concerning the fraud of a trade mark implies that the multitude of marks found on a product's package were formed at different stages of the products' life-cycle, and that these marks were used to communicate meaning throughout the consumer chain. Four men were accused of unlawfully conspiring to use a trade mark of two firms, Bailey & Co and Atkinson, with the view to use the marks to defraud the wholesaler chemist Francis Newbery. The witnesses included Edward Roberts, a glass-bottle maker; William-Henry Greggs, a lithographer printer; Charles Harrison, a bottle-mould maker; and Edward Clark, another bottle-mould maker, emphasising the number of market participants involved in the making and dissemination of packaging. The process of making the package and associated branding would occur at varying locations; a statement from one of the accused men was recounted in court, underlining that "his moulds for making A. Atkinson's and Bailey's bottles would be found at Roberts's, 86, Nicholas Street, New North Road, and that Glindon, of 86, Westmoreland Place, City Road, made the stoppers for him; that Bailey's larger bottles he could get anywhere, and had no mould" (Old Bailey Proceedings Online 2016). The practical arrangements for the creation of a container could be extensive. The case also indicates that trade marks were considered a powerful mediator of trust. Edward Pickering, the chief clerk of Francis Newbury & Sons stated that "I believed the White Rose was Atkinson's preparation and that the Ess Bouquet was Bailey's, or we should not have bought them—I thought their trade mark was on every bottle," and remarked that "they were sold to our customers as genuine articles, we ourselves believing them to be genuine" (Old Bailey Proceedings Online 2016). Pickering's belief that the products were genuine rested on the fact that the articles carried the expected trade marks of the two firms.

Broadening our understanding of paratext to include the materiality of the product's packaging allows us to address significant processes of branding beyond the text, and to start thinking about the product's processes of production, from the layout, illustrations, design practice, and manufacturer of the container. Such an approach allows us to consider how trust in a product was formed at all stages of consumption throughout the complex chain of market participants.

CONCLUSION

Pharmaceutical goods as encased within a package and provided with a label are imposed with a certain social value. A remedy within a package or with a label can find meaning within a legal and economic context. Today, expectations regarding drug efficacy are predominately equated with trust in techno-scientific knowledge, and these expectations are imposed from the presence of paratext found on labels and packaging. Paratext on packages today provides the consumer with safety information about the preparation, with lists of active ingredients, known efficacy, dosage instructions, and a list of side-effects (MHRA 2015): in fact "the packaging, the trademark, and the instructions allow the drug to speak for itself" (Henkel 2004: 37). The latter quarter of the nineteenth century into the early twentieth century was a period when the paratext of healthcare products was being shaped and re-shaped alongside increasing regulation in a legal and economic context. The use of trade marks on packaging, particularly in the wake of the 1875 Trade Marks Act, to help identify the product, to exclude competitors, and protect the market, shows how paratext was employed to help construct the credibility and value of medicine products within a commercial context. The market could also be fiscally regulated via paratext, as evidenced by the medicine stamp. This fiscal paratext had repercussions for how the materiality of packaging was constructed in the nineteenth-century medical marketplace for those products that were liable for duty. The appropriation of this mark to suggest government backing by some proprietors forced the Inland Revenue to alter the text to "No Government Guarantee" to ensure that this wider value could not be attached to the stamp. Paratexts were thus able to communicate value beyond the text.

The increased use of packaging as a medium for branding has been shown via the proliferation of advertisements depicting packaging in *The Chemist and Druggist*, a chief mouthpiece for the trade. These packages were made up of separate parts by different participants, as evidenced by the Old Bailey Court Case, suggesting that branding on chemists' products was more complex than at face value. Closer reading of pharmaceutical packaging reveals the multiplicity of market participants. Addressing these marks as important paratexts demonstrates how participants identified genuine products in the marketplace, and suggests that these were important mechanisms for how participants communicated credibility and authority throughout a depersonalised supply chain. This has implications for scholarly understanding of the nineteenth century marketplace where explanatory models have dismissed consumers as gullible fools duped by patent medicine vendors (Young 1961). Marketing efforts by wholesale and manufacturing firms suggest that they considered their consumer more adept at decision-making than traditional narrative has allowed. The conceptualisation of packaging as a paratext, including its constituent parts, such as logos and other graphic communication, as well as the materiality and shape of the package itself, has wider connotations for assessing consumption practices in the period, opening up discussions about how information surrounding healthcare products could be accessed and used by different strata of society and how value was communicated in a society that still included illiterate and semi-literate consumers. An in-depth study of paratext used in the marketing of healthcare products will give a more nuanced understanding of the economic context of these products in the long nineteenth century, enabling us to survey how they were portrayed, interpreted, sold and ultimately understood.

NOTES

1. Stuart Anderson, interview with Alan Garrett. 1995. "Oral Histories of Community Pharmacy" National Sound Archive, © British Library reference C816/05, Tape 1, Side A, 13.32 mins.
2. Historians of marketing have identified a shift in the latter half of the nineteenth century towards a sales-orientated market, where competition gave rise to product-differentiation and the need to advertise. See Church (2000).
3. The medicine stamp tax lasted from 1783 until 1941. In 1864, the Inland Revenue Office replaced the Stamp Office, changing the text on the label.
4. The 1868 Pharmacy Act established a poisons register for 15 scheduled poisons, and was controlled by the Pharmaceutical Society. The packages for these preparations had to be labelled as a poison, alongside the name and address of the retailer.
5. Branding is defined as material marks as well as imaginary constructs. J. L. Basford holds a broad view of branding, which extends beyond proprietary

branding; this includes marks of production, marks of the state, and marks of institutions. This chapter will similarly hold that all marks found on the material package or within advertisements can be considered as forms of branding. See "Introduction" in Basford (2012).

6. This includes proprietary medicines, surgical instruments, dental preparations, electromagnetic devices, curative foods, tonics, vermin killer, and medical toiletries.

7. Wellcome Library Online (https://archive.org/details/chemistanddruggist, version dated 31 Dec. 2014), searched for *The Chemist and Druggist* between 1875 and 1920, see advertisements from issues.

8. The data collection uses one issue of *The Chemist and Druggist* taken from the months April and October from the years 1880, 1885, 1890, 1895, 1900, and 1905. Using two issues from a given year offsets anomalous results. Where advertisements have not been found in the digitised version, I have found the closest issue available with advertisements; the year 1905 only had one available issue with full advertisements, with the closest comparable issue available being October 1903. See: Wellcome Library Online (https://archive.org/details/chemistanddruggist, version dated 31 Dec 2014).

9. The Patents, Designs, and Trade Marks Act, 1883: The definition of the Act was extended to include, fancy words, brands, and single letters in the case of old marks. There was a reduction in costs of registration; Merchandise Marks Act, 1887: the Act made it illegal for companies to falsely claim that they have a Royal Warrant, "or arms so nearly resembling the same as to be calculated to deceive"; The Trade Marks Act, 1905: A statutory definition of the trade mark was given for the first time, as "a mark used on or in connection with goods, for the purpose of indicating that they are the goods of the proprietor of such mark by virtue of manufacture, selection, certification, dealing with, or offering for sale."

10. The fact that this product announced that it contained "no Opium, Morphia, Calomel, or other poisonous drug", shows there was a commercial advantage in directly noting this to the consumer: for the effects of poison legislation on the paratext of labels, see Anderson and Berridge (2000: 27).

Bibliography

Anderson, Stuart. 2005. *Making Medicines*. London: Pharmaceutical Press.

Anderson, Stuart, and Virginia Berridge. 2000. Opium in 20th-Century Britain: Pharmacists, Regulation and the People. *Addiction History* 95 (1): 23–36.

Barker, Hannah. 2009. Medical Advertising and Trust in Late Georgian England. *Urban History* 36 (3): 379–398.

Basford, Jennifer Louise. 2012. *'A Commodity of Good Names': The Branding of Products, c.1650–1900*. PhD thesis, University of York.

Bently, Lionel. 2008. The Making of Modern Trade Marks Law: The Construction of the Legal Concept of Trade Mark (1860–1880). In *Trade Marks and Brands: An Interdisciplinary Critique*, ed. Lionel Bently, Jennifer Davies, and Jane C. Ginsburg, 3–41. Cambridge: Cambridge University Press.

———. 2014. The First Trade Mark Case at Common Law? The Story of Singleton V. Bolton (1783). *U.C. Davis Law Review* 47 (3): 969–1013.

Brown, P.S. 1976. Medicines Advertised in Eighteenth Century Bath Newspapers. *Medical History* 20 (2): 152–168.

Church, Roy. 2000. Advertising Consumer Goods in Nineteenth-Century Britain: Reinterpretations. *The Economic History Review New Series* 53 (4): 621–645.

Cody, Lisa Forman. 1999. "No Cure, no Money", or the Invisible Hand of Quackery: The Language of Commerce, Credit, and Cash in Eighteenth-Century British Medical Advertisements. *Studies in Eighteenth-Century Culture* 28: 103–130.

Doherty, Francis. 1992. *A Study in Eighteenth-Century Advertising Methods: The Anodyne Necklace*. New York: The Edwin Mellen Press.

Haller, John S., Jr. 1975. The Use and Abuse of Tartar Emetic in the 19th Century Materia Medica. *Bulletin of the History of Medicine* 49 (2): 235–257.

Henkel, Anna. 2004. Drugs in Modern Society: Analysing Polycontextual Things Under the Condition of Functional Differentiation. In *Systems Theory and the Sociology of Health and Illness: Observing Healthcare*, ed. Morten Knudson and Werner Vogd. London: Routledge.

Jones, Claire L. 2013. *The Medical Trade Catalogue in Britain, 1870–1914*. London: Pickering & Chatto Publishers.

Marland, Hilary. 2006. "The Doctor's Shop"– The Rise of the Chemist and Druggist in Nineteenth-Century Manufacturing Districts. In *From Physick to Pharmacology. Five Hundred Years of British Drug Retailing*, ed. Louise Hill Curth, 79–104. Aldershot: Ashgate.

Matthews, Leslie. 1986. The Medicine Stamp Acts of Great Britain. *Pharmaceutical Historian* 16 (1): 2–5.

Price, Roger, and Fraser Swift. 1988. *Catalogue of Nineteenth Century Medical Trade Marks 1800–1880*. London: Science Museum.

Styles, John. 2000. Product Innovation in Early Modern London. *Past and Present* 168: 124–168.

Whorton, James C. 2011. *The Arsenic Century: How Victorian Britain was Poisoned at Home, Work, & Play*. London: Oxford University Press.

Wilkins, Mira. 1992. The Neglected Intangible Asset: The Influence of the Trade Mark on the Rise of Modern Corporation. *Business History* 34: 66–95.

Young, James Harvey. 1961. *Toadstool Millionaires: A Social History of Patent Medicines in America Before Federal Regulation*. Princeton: Princeton University Press.

Newspapers and Periodicals

The Chemist and Druggist. 1875a. Notes and Queries. December 15.
———. 1875b. Notes and Queries. July 15.
———. 1885. Patent Medicine Stamps. December 15.
———. 1890. Squire and Sons Advertisement. January 4.
———. 1900. Rendell Advertisement. July 7.
———. 1910. Thomas and William Henry Advertisement. March 12.
The Medical Press & Circular. 1880. Arthur Cox Advertisement. April 14.
The Medical Times and Gazette. 1875. Stamp Duty. December 11.
The Times. 1884. The Pharmaceutical Conference August 13. The Times Digital Archive: Accessed 12 Mar. 2015.

Archives

Boots Archive. The Boots Company plc. Nottingham.
'Oral Histories of Community Pharmacy' (1995) Interviewer: Stuart Anderson. National Sound Archive at the © British Library. London.

Websites

Old Bailey Proceedings Online. 2016. 5th April, 1886, Trial of Samuel Alker Freeman, Charles Lever, Morgan Edward Williams, and William Edwin Boyes (ref: t18860405–413). www.oldbaileyonline.org. Accessed May 2017.
The Medicines and Healthcare products Regulatory Agency. 2015. Best Practice Guidance on the Labelling and Packaging of Medicine. Accessed May 2017. https://www.gov.uk/government/uploads/system/uploads/attachment_data/file/474366/Best_practice_guidance_labelling_and_packaging_of_medicines.pdf
Wellcome Library Online. 2014. Searched for *The Chemist and Druggist* Between 1875–1920. https://archive.org/details/chemistanddruggist. Accessed May 2017.

PART II

Authority, Access, and Dissemination

CHAPTER 6

Remedies for Despair: Considering Mental Health in Late Medieval England

Natalie Calder

A certayne parson ther was that by þe space of .ii. yeres & more was troubelyd wyth suche mocions [of despair] & often tymes by water & by lande. he had mocions to desperacion & to the moste abhomynable that may be to destroy hym selfe. bothe by water & otherwise. And the person neuer ceasyd. but callyd to god dayly and parseuerantly to strength hym. & euer confessing hymselfe the saruant & worshypper of the holy name of Jesu. [...] And þat was this yt was gyuen hym by a light. that when so euer suche mocion cam to his mynde. that he sulde take that mocion for an occasion & remembraunce to honour the passion of crist & the blyssed vyrgyn hys mother.[1]

My thanks are due to Dr Stephen Kelly for his meticulous feedback on an early draft of this chapter. I use in this chapter the phrase 'late medieval' very loosely to denote *c*.1370–1530, setting aside periodising narratives of the Middle Ages that might restrictively class the writers I discuss as 'post-medieval'. Examining the materials in this way allows us to see how such writers based their considerations of mental disorder on texts written firmly 'within' the Middle Ages.

N. Calder (✉)
School of Arts, English and Languages, Queen's University Belfast,
Belfast, Northern Ireland

© The Author(s) 2018
H. C. Tweed, D. G. Scott (eds.), *Medical Paratexts from Medieval to Modern*, Palgrave Studies in Literature, Science and Medicine,
https://doi.org/10.1007/978-3-319-73426-2_6

The quotation above—depicting the beginning of a short exemplum of a man afflicted by suicidal thoughts—appears in the fifteenth chapter of a text entitled *A deuote treatyse for them that ben tymorouse and fearefull in conscience*, written in 1527 by William Bonde, a brother of the Bridgettine Syon Abbey. The protagonist of the exemplum experiences 'mocions to desperacion'—impulses or inclinations[2]—that result in 'the moste abhomynable [thoughts] that may be to destroy hym selfe'. Bonde pays particular attention to the afflicted man's desperation for relief from his torment; despite his troubled thoughts causing him to consider death as an escape, he continually and without fail turns to his faith to alleviate his despair. After two years of torment, God relieves his suffering, 'gyuen hym by a light, that when so euer suche mocion cam to his mynde', he might turn his darkened thoughts to Christ's Passion instead of his own pain. This is the crux of Bonde's short exemplum: the intensity of despair and temptation to suicide can only be alleviated in this narrative by the afflicted turning back to his faith repeatedly, calling 'dayly and parseuerantly' to God to give him strength enough to resist temptation.

Bonde's treatise focuses centrally on despair, identifying its root causes—scrupulosity, erroneous consciences, a predisposition to melancholy, among others—and providing support to a readership that is struggling to cope with mental disorder brought on by misconceptions of God's grace. The need for such a treatise, Bonde tells us in the first few chapters, stems from confusion surrounding appropriate and inappropriate types of 'fear' in Christianity; indeed, Bonde explains that 'This I write by cause many be deceyued in feryng & dredynge god. And specially som religious parsonis takynge the one feare for the other weneyng that they render to god the dew feare and holy reuerence and yt is not so'.[3] Bonde's treatise seeks to remedy the consequences of such confusion; by illuminating what he sees as the causes of such despair, providing exempla such as the one opening this chapter, and suggesting ways to overcome its afflictions, the sixteenth-century writer composed a text that is specifically designed to care for the mental health of his readers, as much as pastoral materials might be designed to care for the soul.

Initially written at the request of a sister of Syon, Bonde's treatise was then published and distributed among a much wider lay audience in 1527, with a second edition following posthumously in 1534. The copy of the text examined here (STC 3275) is a second edition of Bonde's treatise; its 28 leaves render the treatise comparatively shorter than other works produced by Bonde (most notably his *Pilgrymage of Perfeccyon* at 308 leaves).

The treatise opens with a decorative title page that alerts the reader to the fact that this text forms part of the lucrative vernacular print output from Syon Abbey in the opening decades of the sixteenth century: 'compyled by a Brother of Syon called Wyllyam Bonde, a Bachelar of Diuinite'.[4] The edition contains a brief table of contents, outlining the subject matters of all twenty chapters, and then begins the treatise proper, signalling the opening chapter with a decorative initial. Handwritten annotations throughout the text are few and far between (although it may be that they are undetectable in EEBO's microfilm scans). The text consists of clear titles for each chapter, with catchwords and quire markers appearing at the bottom of the folios. Very brief commentary has been added (in print) throughout the text; such commentary alternates between short biblical references in Latin and signposts to key exempla or explanations which are demonstrative of each chapter's specific topic. This edition of the treatise closes with three woodcut illustrations; the final illustration, depicting the Passion with a short prayer inscribed at the base of the image, is also found in another (rearranged) print of the text (STC/3276). Circulating among an audience with growing exposure to theological writings in the vernacular, *A deuote treatise* found its niche among readers who were encountering complex questions of faith.[5]

The fifteenth and early sixteenth centuries in England saw an outpouring of vernacular religious texts that exposed an increasingly lay audience to diverse and often abstruse methods of contemplation. Providing instruction on how to achieve a 'mixed life'[6]—a spiritually advanced life from without the confines of monastic orders—such texts demanded of their readers an intensity in meditation and self-examination that was potentially difficult to control or manage without a dedicated spiritual advisor. Such theologically complex materials—supplied to an increasingly curious, interrogative and sophisticated laity—engendered fears of misinterpretation and concerns over the potential errors such freedoms could create. They also brought with them the increased likelihood that readers' beliefs would be tested and challenged. The challenges of such exposure to complex questions of belief, coupled with the intense self-interrogation that many advanced contemplative texts demanded, had the potential to result in extreme spiritual despair or 'acedia', an ailment that traditionally was suffered by enclosed religious but, through the vernacular book trade in late medieval England, found articulations (often as 'wanhope') among lay audiences.[7] Such despair was entangled with the conviction that salvation was unachievable, leading the individual to deny the grace and benevolence of God.

The brothers of Syon Abbey were key players in the vernacular book trade of late medieval London: the vast library at Syon, coupled with the lucrative print output and extensive connections the brothers maintained with the lay community surrounding the monastery, attest to its influence on the religio-literary culture of complex vernacular theology in late medieval London.[8] As E. A. Jones and Alexandra Walsham have pointed out, Syon was often overlooked in historiographical accounts of pre-Reformation England, accounts which favour 'a narrative that characterises late medieval monasticism as stagnant, corrupt and moribund, desperately out of touch with the spiritual needs of the wider populace surrounding it' (Jones and Walsham 2010: 9). The print culture generated around Syon Abbey, however, is indicative of a vibrancy in late medieval vernacular religion that broadened the potential for increasingly critical engagement with faith.[9] Such exposure to complicated and, at times, obscure devotional writings led to increasing concern over the potential mistakes in interpretation that could be made. The detrimental effects such confusion could have on those who had little devotional training or spiritual guidance were manifested in cases of despair such as that which Bonde identifies in *A deuote treatyse*.

The present chapter will therefore take as its focus the detrimental effects of the circulation of such theologically advanced books on the mental health of their lay users, understood via Bonde's *deuote treatise*. Considering Syon Abbey's influence on late medieval London, I will focus in particular on William Bonde as one of three prolific brothers of Syon, who were active in the early years of Henry VIII's reign. John Fewterer and Richard Whitford were Bonde's peers at Syon in the first quarter of the sixteenth century, producing a diverse range of vernacular texts for circulation among the wider audience that lay beyond the cloisters' walls.[10] Through *A deuote treatyse*, Bonde sought to provide a remedy for the 'scrupulosity' of readers and practitioners of complex meditative exercises, outlining specifically how such contemplation can cause deterioration in mental faculties and health in general. Perhaps most interestingly, Bonde draws upon (among others) a fourteenth-century writer for the basis of his remedies. William Flete's *Remediis contra temptaciones* was written, like Bonde's later text, at the behest of a nun who was under Flete's spiritual guidance. Flete's text was then translated multiple times, finding traction among vernacular audiences throughout the fifteenth century (albeit not under his name). Texts such as those examined here can be said to form a paratext to mental health concerns in the larger devotional contexts of late medieval England. In her book *Reforming Printing: Syon Abbey's Defence of Orthodoxy 1525–1534*, Alexandra da Costa outlines the different vernacular translations of devotional materials made available by the brothers of Syon: Bonde's treatise on conscience was published alongside 'Whitford's translations of the *Rule of St Augustine* (1525), the Syon *Martiloge* (1526), St Bernard's *Golden Epistle* (1530), and Fewterer's translation of Ulrich Pinder's *Speculum Passionis* (1534)' (da Costa 2012: 31). Bonde's treatise is dubbed one of the 'basic works of spiritual comfort' by da Costa (2012: 31); it seems to me that the text contributes to a much more sophisticated paratext to devotional materials that has as its focus concerns around mental health. Such advice arguably constitutes a form of psychiatric epitext that encodes concern over mental illness within the larger textual landscape of late medieval hermeneutics. I will argue that Bonde's treatise and Flete's earlier *Remediis* provide evidence for sustained and continued concern over the mental health of readers who were exposed to the complexities of late medieval vernacular theology.

Syon Abbey and (Very) Late Medieval Print Culture

Early sixteenth-century England saw an explosion of devotional materials promulgated to broader audiences than had been possible before the advent of new print technologies. Such exposure to sophisticated vernacular theology, a mode of devotion that often promoted a focus on interiority and self-scrutiny, led to the production of texts (such as Bonde's *A deuote treatyse*) which addressed the cognitive challenges of faith.[11] Through the patronage of Henry VIII's grandmother, Lady Margaret Beaufort, printing presses were producing complex religious texts on commission, and disseminating them to a broader audience than before.[12] Susan Powell explains that the circulation of devotional texts via print 'made them accessible as ways of living to a much wider audience, so that the modes of thought in the female enclosed communities of the fifteenth century [such as Syon Abbey] became a model for both women and men in the active life'.[13] The distribution of such devotional texts via the medium of print to a broader audience than had previously been possible brought access for lay readers to complex devotional materials. Syon Abbey became a centralised site of print culture in the opening decades of the sixteenth century.[14] The innovative, and industrious, nature of print culture in which Syon was involved meant that Bonde's duties of care were not limited solely to the sisters of the Abbey. Indeed, by the early sixteenth century, Syon was catering to the sophisticated lay appetite for spiritually and devotionally challenging materials—books that would allow readers to access advanced methods of contemplation—that was taking hold throughout the 'long' fifteenth century.[15] The brothers of Syon maintained a network of connections—both commercial and pastoral—even after entering the Abbey, using their pastoral training and influence to educate and provide guidance for those beyond their immediate care.[16] The sheer output of devotional material for an increasingly lay readership—especially that which was designed, at initial conception at least, for the instruction of the enclosed sisters of Syon—indicates Syon's response to external demands for detailed spiritual instruction that had meditative and cognitive practices as its focus.

Alexandra da Costa, importantly, reminds us that the circulation of complex devotional texts was not a *new* occurrence in Bonde's time:

> At least half of the printed books enjoyed circulation for some years before they were printed. [...] Seen in this light, the printing of these books continued a tradition established in the 1420s of Syon books written for a specific recipient being intentionally proffered to a wider readership. (da Costa 2012: 37)

According to Da Costa, Bonde, Fewterer and Whitford were continuing a practice of book production for the laity that began centuries before. Such circulation is matched in the revisions and translations of Flete's *De remediis*. Bonde's *A deuote treatyse* formed one of the later links in a chain of texts that addressed despair among the late medieval laity directly. *Acedia* was no longer the privilege of the monasteries: despair was experienced and articulated across the social strata of fifteenth-century England. Moreover it is evident that Syon was responding to the demand for devotional materials that could function as a kind of textual spiritual advisor; the circulation of introspective texts in the vernacular indicates an appetite for materials that would aid meditative practices *outside* of the cloister. Bonde's treatise, in its address to those who suffer from despair caused by scrupulosity, responds to the consequences of distributing advanced vernacular theology to wider audiences who perhaps lack the spiritual guidance such materials originally required. *A deuote treatyse* examines the causes of despair and attempts to embody a paratextual 'medecyne'[17] that may be circulated among the very texts that can cause the illness it seeks to remedy.

Diagnosing Spiritual Despair

In their examination of mental disorders in the writings of the twelfth-century Christian mystic, Hildegard of Bingen, Suzanne M. Phillips and Monique D. Boivin consider the fruitful contribution writers of the Middle Ages can provide to the discussion of mental health in the humanities. They state:

> Medieval writers are usually excluded from conversations about mental disorder because the medieval period is dismissed by psychology and psychiatry texts as focused exclusively on supernatural explanations [...] In fact, medieval writers used a range of explanations for mental illness, including biological, astronomical, social, spiritual and dietary observations. (Phillips and Boivin 2007: 359)

Indeed, the caricature of the Middle Ages as *religiose*—that is, too 'religious' to contribute in a 'meaningful' way to scientific, 'secularised' discussions around how the brain functions (and malfunctions)—hinders such scholarship's ability to read effectively the ways in which writers encountered mental disorder among their readers.[18] Historical accounts of mental illnesses such as depression often assume that mental disorder was only—or at the very least, primarily—understood in the Middle Ages by way of physiological misconceptions.[19] Indeed, the dominant form of describing mental illnesses (which find their post-medieval equivalents in illnesses such as depression) was through the paradigm of the four humours: production of black bile signalled melancholia that would then affect the senses and the ability to reason. As we shall see from Bonde and Flete's texts, however, some writers of the Middle Ages had a much subtler understanding of mental health than has been assumed possible. This is not to suggest that contemporary medicinal accounts of humoral imbalances were overlooked by our writers. Bonde tells us in his tenth chapter that:

> Doctours and specially physyciens determine and say that of a certeyne humoure in the stomacke whether that humoure be melancholy as yt is lyke to be or els a [dull] coloure or rather a blacke fleme. I leaue yt to them that be lernyd in physycke but they say that of suche an humoure there rysyth a blacke & a darke fume to the hed whyche so troubelyth the sensis and hed of man that by thoccasion therof reason is darkenyd.[20]

Acknowledging that his area of expertise is not in the diagnosis of physical symptoms—'I leaue yt to them that be lernyd in physycke'—Bonde explains that the 'blacke & a darke fume' caused by an imbalance of the humours rises specifically to the *head*, causing a 'darkening' of reason. Flete similarly acknowledges physical symptoms of despair in the 'angwyschiouse [matters] þat comen of malice of þe feend, or of yuel disposicion of mannes complexion' (221). Importantly, however, for both writers, physiological illness is not the only contributing factor to despair. Indeed, Bonde only mentions the humours in the tenth chapter of his treatise, having dedicated most of his previous writing space to the more pressing aspects of spiritual despair: that of scrupulosity, consciences in error and, just as Flete mentions in the quotation above, the influence of 'the cursyd aduersary our auncient & goostly enemy'.[21]

In the opening pages of *A deuote treatyse*, Bonde explains the differences between having an appropriate, filial kind of fear of (or reverence for) God, and the inappropriate type of fear characterised by 'scrupulosity', a condition in which one has 'a mystruste of [God's] marcy supposyng for euery tryfel dampnacion'.[22] The crux of the matter, for Bonde, is the 'mystruste' the scrupulous person has of God's mercy because, without trust, belief cannot flourish. Scrupulosity forces the individual to obliterate appropriate practices of faith from their mind: in the haste to cleanse the soul of sin, scrupulosity causes the person to deny the efficacy of God's sacraments, and to be sceptical of his mercy. Scrupulosity for Bonde is inextricably linked to doubt and instability in faith: he warns that 'the mansyon of god may not be fyxed in a scrupulose soule whyche is euer trobelyd & ful of wauerynges and doutfulnes'.[23] It is almost impossible for an individual to detect whether they are afflicted with scrupulosity, as it affects the thought processes that go into determining whether one has repented their sins sufficiently. In his sixth chapter, therefore, Bonde provides a long and detailed definition of scrupulosity. Crucially, Bonde's description attributes scrupulosity to disturbances of the mind: it is an 'vndyscrete dejeccion or trouble of the mynde by thoccasion of ymaginacions of certayne thynges whiche semyth to be contrary'.[24] Scrupulosity is made manifest in the 'ymaginacions' or processes of cognition that confuse the individual into thinking their spiritual practices are inadequate. Such disorders present themselves in 'a mynde vnstable', causing them to imagine things that lead to confusion and instability.[25]

Bonde drew on the works of Jean Gerson for his knowledge on the psychological effects of scrupulosity;[26] in order to apply such knowledge pragmatically, however, to create a practical guide on how to manage anxiety that leads to despair, the sixteenth-century writer was influenced undoubtedly by Flete's *De remediis contra temptaciones*, in particular in his discussions of fear and knowledge about spiritual desperation. Syon Abbey's library possessed at least four Latin copies of the text, and Bonde's directions for managing despair demonstrate undeniable connections to Flete's earlier treatise. William Flete was an English Augustinian friar, born in the first quarter of the fourteenth century.[27] Flete studied to become a master of theology, but took umbrage at the corruption he perceived among his peers. He did not complete the final exercises of his degree; instead, he went into self-imposed exile to the Augustinian monastery at Lecceto, leaving England on 17 July 1359. His treatise, *De remediis contra temptaciones*, was written before his departure (*c.* 1358–59), almost certainly in Latin. Later, he became St Catherine of Siena's spiritual advisor, and died *c.* 1390, having never returned to his homeland. *De remediis* survives in two main Latin versions and three English versions.[28] The text in its various forms circulated widely in fifteenth-century England, and such circulation extended into the sixteenth century with Wynken de Worde's edition in 1519.

Flete does not use the term 'scrupulosity' to describe the anxiety generated by fear of sin. That scrupulosity is what Flete addresses in his text, however, is indisputable. In describing the three methods in particular that the fiend uses—'vndir þe colour of goodnesse' (232)—to deceive man, he states: 'On is this, þat þou3 a creature, man or woman, be neuere soo wel ne so ofte schreuen and in reste of soule, þe fend maketh hem to beleve þat þei ben not wel schreuen, and alle he doth to brynge þe soule to heuynesse' (232). The fiend causes the individual to *believe* differently, to think that their acts of penance are insufficient for salvation, disrupting the 'reste of soule' and bringing them to 'heuynesse': 'this he doth', Flete explains, 'for to drawe here hertis to heuynesse […] and so to tempte hem to dispeir and to bitter þou3tis' (233). Despair is antithetical to belief, for it displaces the trust in God that would allow the individual to realise that their devotional practices are adequate. The fiend distorts belief within the despairing soul enough to challenge the efficacy of God's sacraments. Both Flete and Bonde's discussions of despair find their bases in the function (or malfunction) of the conscience. Indeed, the role of conscience was the topic of much writing produced by Syon Abbey in the late fifteenth and early sixteenth centuries. The introspection the brothers cultivated in their texts appeared not only in those designed initially for the *via contemplativa*; such introspection in fact was incorporated into materials written specifically for lay householders to enable them to teach their families the basic tenets of Christianity. In addition to texts such as Bonde's *A deuote treatyse*, Richard Whitford's *Werke for Householders*, for example, provided basic religious education coupled with allusions to interiority (through discussions of conscience) that could be explored further by the more enthusiastic reader in Bonde or Flete's texts.

UNCONSCIONABLE CONSCIENCES

Flete and Bonde both warn that the conscience is the means by which the devil infiltrates the subconscious, rendering the individual unable to determine truth from falsehoods, which leads, if left untreated, to despair. The concept of conscience has a complicated history—primarily due to its ambivalent introduction into early Christian theology[29]—that led to extensive debates and discussions over the attributes and function of 'conscience'. The foundational ambiguity of the term—combining aspects of the Greek (internal) *synderesis* and the Latin (external) *conscientia*—endures in the malleability of the definitions and uses of conscience throughout the Middle Ages and into the sixteenth and seventeenth centuries. It is an ambiguity that resulted in a complex tool for spiritual examination and reflection that meant different things for different writers. Conscience could represent a moral compass, a record of interiority, a self-regulator, a discomfort, a compulsion to do what is right, or a faculty to *decide* what is right in the first instance. Scott L. Taylor outlines Jean Gerson's interpretation of conscience,[30] explaining that:

> synderesis [for Gerson] is a God-infused predilection or affectivity for higher principles. This capacity, however, can only bear fruit insofar as it recognizes something toward which to be drawn. Hence, Gerson argues that cogitation, meditation, and ultimately, contemplation, are essential, though insufficient, to achieve such union. Only when contemplation, a cognitive act characterized by rest, fulfilment, and stability, takes place, can it produce affectivity which rises above and seizes the divine. (Taylor 2014: 418)

For Gerson, *synderesis* can only be successful when contemplation remains intact or perfect, that is, when it is 'characterized by rest, fulfilment and stability'. Bonde's treatise, importantly, deals with the malfunction of this process: the sixteenth-century text focuses on the aftermath of the conscience in *error*, when contemplation has been performed without the stability necessary to meditate effectively.[31] In *A deuote treatyse*, Bonde warns that conscience can be transformed into the mouthpiece of the devil:

> He [the fiend] wyl also somtime speke to vs in the voyce of oure owne conscience specially when we be darked in our reason & say. My fayth is vnstable and not trew. it is but fayned my hope ys not vpwarde as yt shuld be. I haue no loue to god. I consent to euery syn. & to euery delectacion. I wol despayre how may I loke for heuen. that consentes to euery delectacion. and do not resyst eny temptacion euer I decay and fall from god.[32]

Conscience here is susceptible to usurpation, with the fiend speaking in 'the voice of oure owne conscience', gaining the individual's trust. Such usurpation occurs when we are 'darked in our reason', where faith is coloured by instability and untruthfulness. Bonde delivers his warning in the first person, ventriloquising the fiend's infiltration. He continues:

> Suche thoughtes and innumerable moo he wol speke in oure soule as yt were oure wone conscience that shulde speke & all is but hys crafte and suggestions. Somtyme he wol speke in his owne dyuellyshe parson & cast his most terryble engyns of fere in mannes soule to hys grete vnquietnes.[33]

Bonde recognises in his text that, by infiltrating one's conscience during moments of spiritual weakness, the fiend violates the psychological functions that allow the individual to believe confidently in God. The fiend's interference compromises belief, permitting the 'most terryble engyns of fere' to be implanted in man's soul that will cause 'grete vnquietnes' and unbelief to thrive. The mistrust that the fiend implants in the afflicted man causes him to 'suppos[e] for euery tryfel dampnacion in their erronious conscience'.[34] Flete also articulates the fiend's ability to ventriloquise contrarious thoughts via the individual's compromised conscience:

> if suyche an errour of conscyence made be the enemy seye on to ȝou þat other men feele not þat þat ȝe feele, and þerfore thei kunne not deme ne ȝeue ȝou good remedye þerto, and þerfore ȝe muste folwe ȝoure owne fantasyes, or ellis ȝe þenken þat ȝe schulen be lore, take ȝe non heed of this þouȝt and steryng, ne of no suyche fantasyes þat comen in to ȝoure herte ne charge hem not. But putteth awey all suyche errouris of consciens as faste as thei comen to mende. (234)

Here Flete characterises the fiend's usurpation of conscience as singularity, the sin of placing one's opinions and knowledge above that of anyone else's, including God's: because others cannot feel the despair you feel, they cannot help you remedy your despair; 'ȝe muste folwe ȝoure owne fantasyes', says the fiend, and separate yourself from your advisors. Flete articulates the same fear that can be found in all manner of pastoral materials from the period: despair over the prospect of one's spiritual damnation encourages isolation and separation from the Christian community.

Flete's advice is to 'putteth awey all suyche errouris of consciens as faste as thei comen to mende'. The difficulty in this is apparent. If conscience is the means of determining right from wrong, an internal cognitive process that both enables and forces the individual to work through questions of morality and come to a decision, then it is almost impossible for the individual to determine if his/her conscience is compromised, since this is the faculty that is affected most by despair. Secondly, Flete himself explains that 'þe feend is ful besy to men and women of tendir cons(c)yens': those with a 'tendir' or unstable conscience are more susceptible to the fiend's exploits, and so are less likely to have the capacity to identify their erroneous conscience single-handedly. (234) Drawing on Flete, Bonde asserts that despair renders the individual unable to detect that 'the thynge that troubelyth them ys but a tryfle in yt selfe and sone remouyed yf they wolde put ther good wyll there vnto. vndoubted to the greate encrease of verteu in theyr soules'.[35] Despair skews perspective: the sin that causes the descent into scrupulosity is usually 'but a tryfle in yt self'. A conscience in despair distorts the ability to reason, causing the mind to become troubled by its contrary thoughts. Describing those afflicted by despair, Flete states:

> thei ben so byten in cons(c)yens þat þei kan no while to gydir haue reste in hem self; and alle þis þe fend doth þoruȝ fals dreed and blynd conscyens. But þe remedy of þis temptacion and of all other is þat þei gouerne hem be here confessour, or be some good discret persone, and rule hem fully aftir hym, and not aftir here owne blynde mysrulyd consciens. (234)

Conscience provides the entry-point for the fiend to infect the individual with despair, working from within so that 'þei kan no while to gydir haue reste in hem self'. The fiend blinds conscience, rendering one incapable of determining right from wrong and altering the cognitive processes that allow one to think reasonably about their faith and spirituality. Conscience becomes the site of internalised affliction that causes mental disorder for these texts; the erroneous conscience is a wound that infects the psyche, distorting perception and igniting doubts that sever the connection between God and the afflicted.

Conclusion

Flete's advice in his manual of remedies—which is echoed throughout Gerson and Bonde's works too—is that one must seek external counsel, a voice of reason to counteract the 'fals dreed and blynd conscyens'. This is the impetus behind Bonde's treatise: using Gerson's discussion of scrupulosity, Bonde creates a manual for his readers on how to identify their spiritual anxieties and remedy despair, drawing directly on the format of Flete's *Remediis*. His text embodies the external voice of reason that his wider readership—those without personal access to Bonde's counsel—requires in moments of despair. Both Bonde and Flete attribute the cause of the mental illness they describe in their texts to the malfunction of conscience. The cognitive faculty that is used to make decisions on the efficacy of penance turns against the individual, betraying them to 'fantasies' that do not make sense in a Christian devotional context. Erroneous consciences cause despair because moral 'truth' is negated; there still exists within the individual a desire to align one's beliefs with the external, wider Christian narrative. The inability to reconcile one's beliefs with the moral and spiritual framework that Christianity provides results in a deterioration in health that primarily affects mental faculties. Scrupulosity and despair become signifiers of mental illness that Flete and Bonde both attempt to remedy in their texts. Flete's *Remediis* and Bonde's *Treatise* might seem incongruous choices for a collection of essays on medical paratexts; these texts have as their focus, however, care of the mental health of their readers, acting as an accompaniment to more general manuals concerned with the care of the soul. Flete and Bonde emerge from a tradition that demanded depth in contemplation and intense interrogation of the participants' mental capacities; their remedies, grounded in spiritual concerns, recognise the correlation between physical and mental ailment, and seek to soothe, comfort, and attempt to cure mental distress caused by advanced, difficult and often impossible exercises in spiritual and divine cognition.

Notes

1. Bonde, sig. E2[v].
2. The *MED* entry for 'mocion' defines the term as 'An impulse; inner prompting, inclination; desire, wish', and even possibly 'motive' or an 'instigator'. The term is not to be confused with the modern *emotion*.
3. Bonde, sig. A4[v].
4. Another type-set of the text (STC 3276) includes additional paratextual detail: the treatise was 'sent to a deuote Relygiouse woman of Denssey. At þe instance of one of her sprituall frendes. And by that same frende ouerseen and deuyded in to xx. Chapitres to the more comforte of the Reders'. This additional editorial involvement, occurring after the death of Bonde, has also provided an alternative title for the text: 'the Consolatori of Timorouse and fearefull consciencys'.
5. For an excellent account of the religious vernacular book culture in England at this time, see Bose (2005).
6. See, for example, Walter Hilton's *Mixed Life*.
7. See Wenzel (1966) and, more recently, Crislip (2005).
8. See Powell (2010) and Rhodes (1993a).
9. See Gillespie (2001).
10. For more on Fewterer, see Rhodes (1993b); for more on Whitford, see Appleford (2016).
11. See especially Hellinga (2010).
12. For information on the life of Lady Margaret Beaufort, see Jones and Underwood (2004).
13. Powell (2011) provides a very useful overview of the advanced materials that were printed and distributed in this period.
14. For more information, see Powell (2010) and Gillespie (2005).
15. See especially Kelly and Perry (2011).
16. Da Costa (2012) has examined Syon's outpouring of religious texts, with specific reference to Bonde, Whitford and Fewterer.
17. Sig. E1[v].
18. For an excellent corrective to such periodising and secularising narratives, see Davis (2008).
19. See, for example, Babb (1951) and, more recently, Varga (2013).
20. Sig. C2[r].
21. Sig. D4[r].
22. Sig. A3[r].
23. Sig. B1[r].
24. Sig. B2[r].
25. It is possible that such 'ymaginacions' could be read as the malfunction of meditative techniques provided in vernacular devotional materials. More work (outside the scope of this chapter) needs to be done on this prospect.

26. See Brown (1987).
27. Detailed accounts of Flete's life can be found in: Colledge and Chadwick (1968), Hackett (1992) and Webb (2004) (a brief synopsis is given here, based on these sources).
28. For details on the textual transmission of *De remediis*, see Hackett et al. (1964).
29. For a brief overview of conscience from its origins to its uses today, see Strohm (2011) and Potts (1980).
30. This is important to our discussion because Gerson is Bonde's most-cited authority in his treatise.
31. See Kolnai (1957–58), an article which makes the distinction between 'overlain consciences'—i.e. those that accept a different moral framework (as is the case in heresy) and 'erroneous consciences' as consciences that malfunction.
32. Sig. F1^{r-v}.
33. Sig. F1^v.
34. Sig. A3^r.
35. Sig. B1^v–B2^r.

BIBLIOGRAPHY

Primary Sources

Bonde, William. 1534. *A deuote treatyse for them that ben tymorouse and fearefull in conscience*. London: M. Fawkes. STC 3275 and STC 3276.
Flete, William. 1968. In *Remedies Against Temptation: The Third English Version of William Flete*, Archivio italiano per la storia della pietà, ed. Edmund Colledge and Noel Chadwick, vol. 5. Rome: Edizioni di storia e letteratura.
Hilton, Walter. 1986. In *Mixed Life Edited from Lambeth Palace MS 472*, ed. S.J. Ogilvie-Thomson. Salzburg: Institut für Anglistik und Amerikanistik.
Whitford, Richard. 1537. *A Dayly Exercise and Experience of Dethe*. London: Johan Waylande. STC 25414.

Secondary Sources

Appleford, Amy. 2016. Asceticism, Dissent, and the Tudor State: Richard Whitford's Rule for Lay Householders. *Journal of Medieval and Early Modern Studies* 46: 381–404.
Babb, Lawrence. 1951. *The Elizabethan Malady: A Study of Melancholia in English Literature from 1580 to 1642*. East Lansing: Michigan State College Press.
Bose, Mishtooni. 2005. Vernacular Philosophy and the Making of Orthodoxy in the Fifteenth Century. In *New Medieval Literatures*, ed. Wendy Scase, Rita Copeland, and David Lawton, vol. 7, 73–99. Oxford: Oxford University Press.

Brown, D. Catherine. 1987. *Pastor and Laity in the Theology of Jean Gerson.* Cambridge: Cambridge University Press.

Colledge, Edmund, and Noel Chadwick. 1968. Introduction. In *Remedies Against Temptation: The Third English Version of William Flete*, Archivo italiano per la storia della pietà, ed. Edmund Colledge and Noel Chadwick, vol. 5, 199–240. Rome: Edizioni di storia e letteratura.

Crislip, Andrew. 2005. The Sin of Sloth or the Illness of the Demons? The Demon of Acedia in Early Christian Monasticism. *The Harvard Theological Review* 98: 143–169.

Da Costa, Alexandra. 2012. *Reforming Printing: Syon Abbey's Defence of Orthodoxy 1525–1534.* Oxford: Oxford University Press.

Davis, Kathleen. 2008. *Periodization and Sovereignty: How Ideas of Feudalism and Secularization Govern the Politics of Time.* Philadelphia: University of Pennsylvania Press.

Gillespie, Vincent. 2001. *Syon Abbey*, Corpus of British Medieval Library Catalogues. Vol. 9. London: British Library.

———. 2005. Syon and the English Market for Continental Printed Books: The Incunable Phase. *Religion and Literature* 37: 27–49.

Hackett, Benedict. 1992. In *William Flete, O. S. A., and Catherine of Siena: Masters of Fourteenth Century Spirituality*, The Augustinian Series, ed. John E. Rotelle, vol. 15. Villanova: Augustinian Press.

Hackett, Benedict, Eric Colledge, and Noel Chadwick. 1964. William Flete's "De remediis contra Temptaciones" in Its Latin and English Recensions: The Growth of a Text. *Mediaeval Studies* 26: 212–230.

Hellinga, Lotte. 2010. *William Caxton and Early Printing in England.* London: British Library.

Jones, Michael K., and Malcolm G. Underwood. 2004. Beaufort, Margaret, Countess of Richmond and Derby (1443–1509), *Oxford Dictionary of National Biography.* Oxford: Oxford University Press. http://www.oxforddnb.com/view/article/1863. Accessed 1 Oct 2016.

Jones, E.A., and Alexandra Walsham. 2010. Introduction: Syon Abbey and Its Books: Origins, Influences and Transitions. In *Syon Abbey and Its Books: Reading, Writing and Religion, c.1400–1700*, ed. E.A. Jones and Alexandra Walsham, 1–38. Woodbridge: Boydell Press.

Kelly, Stephen, and Ryan Perry. 2011. Devotional Cosmopolitanism in Fifteenth-Century England. In *After Arundel: Religious Writing in Fifteenth-Century England*, 363–380. Turnhout: Brepols.

Kolnai, Aurel. 1957–58. Erroneous Conscience, *Proceedings of the Aristotelian Society*, 58, 171–198.

Middle English Dictionary. University of Michigan, 2014. https://quod.lib.umich.edu/cgi/m/mec/med-idx?type=byte&byte=117233968&egdisplay=open&egs=117252023. Accessed 31 July 2017.

Phillips, Suzanne M., and Monique D. Boivin. 2007. Medieval Holism: Hildegard of Bingen on Mental Disorder. *Philosophy, Psychiatry & Psychology* 14: 359–368.

Potts, Timothy C. 1980. *Conscience in Medieval Philosophy*. Cambridge: Cambridge University Press.

Powell, Susan, 2010. Syon Abbey as a Centre for Text Production, In *Saint Birgitta, Syon and Vadstena*, ed. Claes Gerjrot, Sara Risberg, and Mia Åkestam, 50–70. Konferenser: 73. Stockholm: Kungl. Vitterhetsakademien och anti-kvitets akademien.

———. 2011. After Arundel But Before Luther: The First Half-Century of Print. In *After Arundel: Religious Writing in Fifteenth-Century England*, 523–541. Turnhout: Brepols.

Rhodes, J.T. 1993a. Syon Abbey and Its Religious Publications in the Sixteenth Century. *Journal of Ecclesiastical History* 44: 11–25.

———. 1993b. Prayers of the Passion: From Jordanus of Quedlinberg to John Fewterer of Syon. *Durham University Journal* 85: 27–38.

Strohm, Paul. 2011. *Conscience: A Very Short Introduction*. Oxford: Oxford University Press.

Taylor, Scott L. 2014. Affectus secundam scientiam: Cognitio experimentalis and Jean Gerson's Psychology of the Whole Person. In *Fundamentals of Medieval and Early Modern Culture: Mental Health, Spirituality, and Religion in the Middle Ages and Early Modern Age*, ed. Albrecht Classen, 406–423. Berlin: De Gruyter.

Varga, Somogy. 2013. From Melancholia to Depression: Ideas on a Possible Continuity. *Philosophy, Psychiatry & Psychology* 20: 141–155.

Webb, Diana. 2004. Flete, William, (fl. 1352–1380), *Oxford Dictionary of National Biography*. Oxford: Oxford University Press. http://www.oxforddnb.com/view/article/53636. Accessed 14 Sept 2016.

Wenzel, Siegfried. 1966. "Acedia" 700–1200. *Traditio* 22: 73–102.

CHAPTER 7

The Medical Paratext as a Voice in the Patient's Chamber: Speech and Print in *Physick for the Poor* (1657)

Elspeth Jajdelska

INTRODUCTION

In the last fifteen to twenty years, the history of medicine and the history of the book have fruitfully coincided to explore the idea of early modern medical provision as a marketplace. The status of books as consumer goods, appealing to diverse segments of the population, can make them valuable indices of those consumer patients themselves, while the authors of medical books can be useful proxies for the different kinds of practitioner competing for those patients' custom.[1] In this respect, medical texts from this period are exhibits in the material and economic history of the time. But in this chapter I show an additional and complementary use of these texts by paying attention to more general findings on the history of paratexts.

While the production, sale and consumption of books have all left material traces in the historical record, the ephemeral speech and social encounters between practitioner and patient have not, and this is especially true of encounters with the poorest patients. But with a new understanding

E. Jajdelska (✉)
School of Humanities, Strathclyde University, Glasgow, UK

© The Author(s) 2018
H. C. Tweed, D. G. Scott (eds.), *Medical Paratexts from Medieval to Modern*, Palgrave Studies in Literature, Science and Medicine, https://doi.org/10.1007/978-3-319-73426-2_7

of the relationship between paratexts and the spoken word in the period, I will argue, we can recreate something of those spoken encounters. In what follows, I first cover recent findings on the relationship between speech, print and paratexts in general in the late seventeenth century. I show that paratexts to religious books aimed at poor audiences recreate a social setting in which elite peers of the author are conceived of as observing the practitioner address the poor and assessing his (pronoun used advisedly) competence. I then go on to analyse the paratexts to a medical text addressed to a poor audience and illustrate how this supports recent findings, which suggest that relationships between practitioners themselves, as well as between patients, could be fluid and ephemeral, shifting between alliances and competition in response to the wide range of social settings in which medicine was practised.

SPEECH, PRINT AND PARATEXTS IN THE LATE SEVENTEENTH AND EARLY EIGHTEENTH CENTURIES

In 1710, Lady Mary Chudleigh published *Essays Upon Several Subjects in Prose and Verse*. It included a reprint of *The Ladies Defence*, a comic verse critique of early modern patriarchy.[2] Her preface complained that two earlier, unauthorised, printings of this poem had respectively '*mangl'd, alter'd, and considerably shortned*' her preface and '*omitted both the* Epistle Dedicatory *and the* Preface' and that this had '*left the* Reader *wholly in the Dark, and expos'd me to Censure*'.[3] What kind of censure could the absence or mangling of a paratext invite? To answer this question, we must consider the way print was conceptualised in relation to speech in the period. In the present day, texts usually make their way from author's computer to reader's hand without being spoken at any point. Alternative textual journeys do exist, through spoken word poetry and audiobooks, for example, but these are the exception rather than the rule. As a rule, print in developed, literate societies constitutes a different category of utterance from speech.

This was not necessarily true in the past. In the late seventeenth century a text could be seen as a proxy for the author's speech (Jajdelska 2016: 57–97). The diarist Samuel Pepys, for example, remarked on George Mackenzie's *Religio Stoici*[4] that, 'In some places he *speaks* well, but generally is but a sorry man' (my italics).[5] His friend John Evelyn also equated personal address by an author with printed or written texts. In the seventeenth century, as now, the verbs 'present' and 'deliver' could be used in at least two senses:

1. speaking
2. handing over a material object

Evelyn tended to conflate these two senses, that is to conflate addressing someone in person with handing that person a printed or written copy of the same text (Jajdelska 2016: 79–82).

This late seventeenth-century conceptual framework meant that publishing was subject to the social norms of speech. At the time these norms included a prohibition on uninvited speech from inferiors to superiors. Gentlemen, for example, were not only expected to 'refrain from blunt or otherwise direct questions' to noblemen, but even to avoid direct eye contact with them (Bryson 1998: 167). Wives, children and servants were to wait for a 'fit time and iust occasion of speech' before they addressed the master of the house, according to the early seventeenth-century clergyman William Gouge.[6]

In addition, all ranks were subject to norms that governed access to knowledge. Speculative thought and public policy, for example, were proper only to men of gentle or noble rank, while household management was proper to women of all ranks (Jajdelska 2016: 114–115). As one irate late seventeenth-century judge put it, 'Every Man that meddles out of his Province is saucy.'[7]

In this light, Chudleigh's concern over mangled or missing paratexts makes sense. Since going into print was equivalent to speaking in public, only the very highest ranks could avoid a breach of decorum if their topic—in this case, verse on the roles of women—was proper to elites. Paratexts were essential to mitigate this impropriety. Chudleigh, for example, uses two favourite defences for addressing superiors without invitation in print: the work was intended for the amusement of friends; and it was printed at their insistence (Jajdelska 2016: 100–111). Indeed, paratexts to late seventeenth- and early eighteenth-century texts recall the ancient origins of the preface: preliminary remarks before a live oration or dramatic performance (see Dunn 1994). The seventeenth-century paratext then can be understood as an analogue for a face-to-face, spoken event in a specific social context. As a result, they can yield valuable clues about those social contexts.

Paratexts and Books Addressed to the Poor

Encounters with the poor leave comparatively few traces on the historical record. Books apparently directed 'at' or 'for' the poor, however, can shed some surprising light on the social and material spaces shared by lowest and highest ranks. The titles of Arthur Dent's *Plaine Mans Path-way to Heauen* (1601) and Richard Baxter's *Poor Man's Family Book* (1674) suggest that these shared spaces hosted a direct address from learned, and therefore comparatively high-ranking, authors to socially inferior hearers or readers.[8] But the paratexts point to the social and material setting of seventeenth-century churches, in which seating and standing areas were organised around social hierarchies. In church, the higher ranks could be either the preacher's default addressees or his evaluators when he turned to instruct the lower ranks. So Baxter's paratexts plead to the preacher's superiors for approval, rather than addressing the poor themselves in their less comfortable quarter of the church (Jajdelska 2016: 104–105).

LAW'S *PHYSICK FOR THE POOR* (1657)

In what follows I analyse the paratexts of a medical text whose title points to a clearly defined group of 'the poor' as addressees, but whose paratexts indicate a more complex speech event. I know of no surviving information about Thomas Law, whose *Naturall Experiments or Physick for the Poor* was published in 1657, beyond his description on the title page as 'Student in Physick'.[9] This looks like a nod to his pioneering and enduringly successful predecessor, Nicholas Culpeper, author of *A Physicall Directory, or, A Translation of the London Dispensatory*, whose later editions describe the author as 'Gent. Student in *Physick* and *Astrologie*.[10] Mary E. Fissell has shown how Culpeper's name dominated the market for medical books for a century, so that authors of popular medical texts with 'links to astrology or chemical medicine' often used or alluded to this title of 'gentleman student of physic and astrology' (Fissell 2007: 115–118).

Culpeper, to their chagrin, translated the College of Physicians' *Pharmacopoeia* into English, with added advice for lay readers on specialist terminology and how to make up recipes, and the book was a staple of lay medicine for more than a hundred years (see Furdell 2002: 42–44; Fissell 2007: 15). Subsequent titles, such as Thomas Cock's *Kitchin-Physick: or Advice to the Poor* of 1676, show a continued association between instruction in physic, and helping the poor by making medicines cheaper and medical theory more accessible.[11]

But a lengthy and rhetorically decorous opening paratext to Culpeper's third edition is addressed not to lay people but to his learned peers and superiors. This address to 'The Colledge of Physjtians' reprimands them for exploiting a monopoly, a reprimand he must make, he claims, because he is their social equal: 'I was born a gentleman and cannot flatter'.[12] He then follows this with an address to 'To the Reader'. This includes more typically decorous pleas to forgive uninvited speech and presumption: 'Praise the book as you find it'; 'Pure pity to you was the motive'.[13] The closest spoken analogues to Culpeper's paratexts then might be first a public, critical, but socially legitimate address to his physician peers, and observed by lay gentry and noblemen who evaluated his performance (for a similar context, see the paratexts to Richard Baxter's *Poor Man's Family Book*).

Law participates in this paratextual culture in a number of ways. The designation 'student of physick' aligns him with Culpeper. But Law, unlike Culpeper, limits his receipts to plants grown in England, making the medicines accessible not just to the poorest, but also to those remote from towns. In doing so, he moves beyond the apparently simple model of institutions (the College of Physicians, the Society of Apothecaries) versus patients to a more complex ecosystem of healers and healing:

> If Latine Authours may be read by some, there are others that cannot read Latine, to such it is as good it had not been writ: or suppose there are enough [receipts] to be bought in English, sure I am that the prices are so high, and the Medicines so chargeable, that nothing can be had without going to the Apothecary, and perhaps give twelve pence for such things as they may have out of a Garden for three pence, and how shall such do; who dwell farr from Apothecaries, (as many Country people who live some six, some eight miles or more from any Citty or Market Town).[14]

Law's account shows that healers could not be neatly divided into physicians, surgeons and apothecaries. Country people, poor people, those who knew no Latin, could, with the help of recipe books (manuscript as well as print) and some cooking equipment, produce medicines for themselves and share them with their neighbours (Leong and Pennell 2007: 135). And, as Alun Withey has shown in a study of health in seventeenth- and eighteenth-century Wales, a given patient might seek advice from a range of sources, including physicians, folk or even magic practitioners, and books (Withey 2011: 57). In rural areas, as Ian Mortimer points out, many patients consulted family and community in the first instance; then perhaps a local apothecary; and finally they would consider a trip to a town to consult an apothecary or physician (Mortimer 2007: 84–85).

Law's paratexts depart from Culpeper's as well as his aims. As we saw earlier, Culpeper's paratexts create a model social context in which he sermonises to one group (the College of Physicians) and then turns to another (gentle- and noblemen, but not physicians) with diffident mitigations of his presumption. This recalls the public disputations in which speakers addressed an opponent while hoping to win the approval of a third party, whether the judge or the audience (Jajdelska 2016: 16–18, 105–106). Culpeper's second paratext bears the marks of the gentleman author addressing this audience uninvited and attempting to mitigate his own impertinence.

Law's paratexts, however, are less clear-cut. His epistle to the reader, like Culpeper's, and like most other authors writing uninvited on elite topics, assumes a learned and potentially hostile audience. For example, he anticipates a reader who will object to the propriety of producing a small book when great ones by his betters already exist: 'Some may say this is a little pamphlet, there are greater Volumes and many Books, both of Physick and Chyrurgery published by men of known abilities.'[15] He assumes, that is, that at least some of his readers are both wealthy (they can afford large books) and learned. Law replies that 'the poor are but little the better' for these more eminent authors, because their works are 'for the most part in the Latine tongue' and that 'the prices are so high, and the Medicines so chargeable, that nothing can be had without going to the Apothecary, and perhaps give twelve pence for such things as they may have out of a Garden for three pence'. How, he adds 'shall such do; who dwell farr from Apothecaries, (as many Country people who live some six, some eight miles or more from any Citty, or Market Town)'.[16] So far, then, Law assumes a gentleman addressee and discusses the poor in the third person. Yet this changes in the next sentence: 'Here, kind Reader thou maist have a bil suitable to thy distemper, for little or nothing, and in thy native tongue ready to serve thee for another time.'[17] The gentleman reader, then, is no longer an observer of Law's efforts on the behalf of the poor, but a beneficiary, whose purchase of the book will bring not just a cheap cure for now, but a recipe to be used again in the future. The envisaged user of the book, though still not the addressee, is in need of Law's 'plainnes, and brevity … that the meanest capacities may reach it, as well with their purses as understandings'. He concludes with the 'kind Reader again, hoping he or she will judge the book 'worth my labour in writing, or thy paines in reading, or that it may be helpful to any one, though the poorest of thy fellow creatures'.[18]

Law's epistle to the reader recalls those of Baxter and Dent in their spiritual guides for the poor. While the poor are to be *beneficiaries* of the work, 'the reader' of the paratext is the author's social equal or superior, and must be appeased as well as persuaded into buying the book and giving or reading it to the poor. However, Law's paratext is also ambiguous; 'the reader' could be a beneficiary as well as a patron, someone who, however learned, will be glad of a useful receipt that can be prepared at low cost. So his learned reader cannot quite be aligned with either Baxter's elite parishioners in the gallery, or Culpeper's potential audience at a disputation.

Similar ambiguities characterise a second paratext in the volume. Law combined his collection of recipes, similar to those given to or used by apothecaries and collected in household volumes, with a second part, *Naturall Experiments*, containing less obviously medical recipes, ranging from 'magical' spells to beauty treatments: '*To know any man or womans minde when they are asleep*'; '*To dye Hair Black*'.[19] In the address 'To the Reader' for this second part, Law explains that the recipes are taken from a manuscript by 'a very ancient Doctor of Physick, (and a friend of mine)', from which he has 'made choice of the best of them, adding some of my own', and that he has published it alongside his '*Poor Mans Physitian*' because the addition would not make the whole book 'much the dearer'.[20]

As with his first epistle, Law assumes a gentleman reader, one authorised to criticise authors who publish (and by proxy speak) on learned topics:

> I believe it [his book] shall be hard spoken of, before it be heard to speak for it self, for thus it hath fared of late, with men of far greater abilities, then my self, by divers who possibly, if the truth were known, are ignorant of the things they carp at.[21]

But Law also suggests that his gentlemen judges are *not* in fact qualified to condemn his work because they lack his specialist learning. When he turns to direct address in the second person, it is with a tone of authority:

> I have bestowed both cost and paines, therefore I shall desire the [sic] to read it thoroughly, and judge indifferently, and if thou likest it, practice understandingly.[22]

As with the first epistle, Law acknowledges two distinct audiences: those entitled by rank to upbraid his presumption in publishing at all; and those who need and acknowledge his learning and expertise.

So while both of Law's paratexts can be accommodated in a traditional paratextual setting (uninvited address to an audience of peers or superiors), they are also consistent with an alternative social context, in which an experienced healer, paid or unpaid, licensed or unlicensed, physician or apothecary, is in the patient's chamber, and providing both recipes and guidance on their usage and preparation. In the recipes themselves, it is this second situation which is brought to mind, again characterised by use of the second person pronoun. Here for example is an excerpt from guidance on how 'To take the perfect draught of any Picture': 'lay on the strong Water as you were shewed before, and any Rowling Presse printer will print it of for you'.[23] The (female?) client here was not 'told' or 'instructed' before, but 'shewed', implying a material context of authority and client in a shared space; most likely a kitchen, where household medicines were typically made; 'Kitchen physic is the best physic', as one proverb had it (Cook 1986: 30).

This characterisation of the paratext's social and material context is consistent with research by Jennifer Richards on how *The Womans Book* was used in early modern England. This was 'one of the most popular midwifery books of early modern England', first published in 1554 and continuously reprinted over the following century (Richards 2015: 435–436). Richards notes that the author, Thomas Raynalde, makes a (paratextually unusual) direct address to women in his 'prologue to the women readers' (Richards 2015: 437). This prologue sketches out a 'scene of reading in the birthing room', which, as Richards observes 'most feminist historians do not' take seriously, but which she suggests we should (Richards 2015: 436). As with Law, however, and (Richards observes) other early modern medical paratexts, Raynalde's first, and in some ways more worrisome, audience, is of men. Some of these men on 'readynge or hearynge' the book 'shal be mooued' to 'abhorre the company of woomen' and to make obscene jokes about women's bodies with one another (Richards 2015: 441–442). This male audience can be conceived of in the anatomy theatre, listening to the lecturer (Richards 2015: 452–454). The setting for the female audience is less clear from Raynalde's text. But Richards provides evidence from illustrations and a comparator text in manuscript that this could be a context in which a female midwife teaches a deputy in the birthing room itself (Richards 2015: 454). She shows, then, that the paratext and text of *The Womans Book* could be conceived of not only as an address by a professor to students in a lecture theatre, but as one female health practitioner addressing another in the birth chamber.

Returning to Law's paratexts, we can now see him turning from a confrontation with his learned peers, to his lower-ranked healers in the sick room. And the group he does not appear to address directly is 'the poor' of his title: *Physick for the Poor*. His audience of healers—who could range from housewives to apothecaries—are the ones who can use his expertise to directly help their poorer, less expert or more remotely situated neighbours:

> I have studied to compose this Method of Physick and Chyrurgery with as much plainnes, and brevity as may be, that the meanest capacities may reach it, aswell [sic] with their purses as understandings, and if (kind Reader) thou Iudgest it worth my labour in writing, or thy paines in reading, or that it may be helpfull to any one, though the poorest of thy fellow creatures return the Praises, and the glory of all unto God, then have I attained the expected end of this my Labour. (Fare well. T.L.[24])

'The meanest capacities' (in the sense of level of intellectual attainment) is not a description that would apply to the learned men of the College of Physicians.[25] But neither can it apply comprehensively to 'the poor', at least some of whom would be unable to read the recipes without help. And, however the author studied to reduce the book's cost by reducing its length, it would still be beyond the reach of a large number of the poor, for whom even a broadside (which cost roughly as much as a loaf of bread) was a significant expense (Marsh 2010: 251). And the 'kind reader' is explicitly distinguished from 'the poorest of thy fellow creatures' by the use of 'thy'. All of this points to an audience of healers, lower in rank than the author, with an altruistic desire to acquire cheap medicines for the poorest. These can then be passed on in their own social networks of neighbours swapping recipes and medicines, or through a cash exchange with an apothecary, or something in between (see Cook 1986: 28–34; Wallis 2007: 59–62; Leong and Pennell 2007: 138–146).

Conclusion

Read in the light of new evidence on speech and print, the paratexts to Thomas Law's *Physick for the Poor* can be read as analogues for real-life social events that incorporated speech, including oral/aural readings of texts. They offer support for recent histories which identify healing as a fluid social practice that moved effortlessly between commercial, local, domestic and institutional spheres. Different kinds of healers could enter temporary alliances, founded variously on charity, profit or social exchange, and collaborate to share expertise and resources as best fitted their own and the patients' needs. In this case, the Law paratexts indicate such an ephemeral partnership between the comparatively expert author ('student of physick') and different makers of medicines, at home or in the apothecary's shop, labouring for the family or for customers, or—as, potentially, in this case—for those of their neighbours in greatest need.

NOTES

1. Mary E. Fissell, 'The marketplace of print', in Mark S.R. Jenner and Patrick Wallis eds. *Medicine and the Market in England and Its Colonies, c. 1450—c. 1850* (Basingstoke: Palgrave Macmillan, 2007) 108–132, 115, 118. Elizabeth Lane Furdell *Publishing and Medicine in Early Modern England* (Rochester, New York: University of Rochester Press, 2002).
2. Mary Chudleigh *Essays Upon Several Subjects in Prose and Verse* (London: T.H. for R. Bonwicke and others, 1710); Margaret Ezell *The Poems and Prose of Mary, Lady Chudleigh* (Oxford: Oxford University Press, 1993), xvii—xxxiv.
3. Chudleigh *Essays*, ECCO images 8–9.
4. George Mackenzie *Religio Stoici* (London: for George Sawbridge, 1663).
5. Samuel Pepys, *The Diary of Samuel Pepys*, eds. Robert C. Latham and William Matthews, 11 volumes (London: 1970–1983). vol. VIII, 10 April 1667, 162.
6. William Gouge *Of Domesticall Duties: Eight Treatises* (London: John Haviland for William Bladen, 1622) EEBO image 152.
7. *A Complete Collection of State-trials, and Proceedings upon High-Treason, and other Crimes and Misdemeanour*, third vol. (London: for J. Walthoe and others, 1730) 2nd ed., ECCO image 879.
8. Richard Baxter, *The Poor Man's Family Book* (London: R.W. for Nevill Simmons, 1674). Arthur Dent, *The Plaine Mans Path-way to Heauen* (London: for Robert Dexter, 1601).
9. Thomas Law *Naturall Experiments or Physick for the Poor* (London: for Edward Farnham, 1657).
10. Nicholas Culpeper *A Physicall Directory, or, A Translation of the London Dispensatory* (London: for Peter Cole, 1649). Nicholas Culpeper *A Physical Directory* (London: for Peter Cole), 1651, 3rd ed., EEBO image 1.
11. Thomas Cock, *Kitchin-Physick: or Advice to the Poor, by way of Dialogue... with rules and directions how to prevent sickness and cure Diseases by Diet* (London: for J.B., 1676).
12. Culpeper *Directory* 1651, 3rd ed., EEBO image 3.

13. Culpeper *Directory* 1651, 3rd ed., EEBO image 5.
14. Law *Physic for the Poor* EEBO image 3.
15. Law *Physick for the Poor* EEBO image 3.
16. Law *Physick for the Poor* EEBO image 3.
17. Law *Physick for the Poor* EEBO image 3.
18. Law *Physick for the Poor* EEBO image 4.
19. Law *Physick for the Poor* EEBO images 65, 70.
20. Law *Physick for the Poor* EEBO images 61–63.
21. Law *Physick for the Poor* EEBO image 62.
22. Law *Physick for the Poor* EEBO image 62.
23. Law *Physick for the Poor* EEBO image 78.
24. Law *Physick for the Poor* EEBO image 4.
25. On early modern meanings of 'capacity', see Jajdelska *Speech, Print and Decorum* 2016, 454.

BIBLIOGRAPHY

A Complete Collection of State-trials, and Proceedings upon High-Treason, and other Crimes and Misdemeanour, third vol. 2nd ed. (London: for J. Walthoe and others, 1730).

Baxter, Richard. 1674. *The Poor Man's Family Book*. London: R.W. for Nevill Simmons.

Bryson, Anna. 1998. *From Courtesy to Civility: Changing Codes of Conduct in Early Modern England*. Oxford: Clarendon Press.

Chudleigh, Mary. 1710. *Essays Upon Several Subjects in Prose and Verse*. London: T.H. for R. Bonwicke and Others.

Cock, Thomas. 1676. *Kitchin-Physick: Or Advice to the Poor, by Way of Dialogue... with Rules and Directions How to Prevent Sickness and Cure Diseases by Diet*. London: for J.B.

Cook, Harold John. 1986. *The Decline of the Old Medical Regime in Stuart London*. Ithaca/London: Cornell University Press.

Culpeper, Nicholas. 1649. *A Physicall Directory, or, a Translation of the London Dispensatory*. London: For Peter Cole.

Dent, Arthur. 1601. *The Plaine Mans Path-Way to Heauen*. London: For Robert Dexter.

Dunn, Kevin. 1994. *Pretexts of Authority: The Rhetoric of Authorship in the Renaissance Preface*. Stanford: Stanford University Press.

Ezell, Margaret. 1993. *The Poems and Prose of Mary, Lady Chudleigh*. Oxford: Oxford University Press.

Fissell, Mary E. 2007. The Marketplace of Print. In *Medicine and the Market in England and Its Colonies, c. 1450–c. 1850*, ed. Mark S.R. Jenner and Patrick Wallis, 108–132. Basingstoke: Palgrave Macmillan.

Furdell, Elizabeth Lane. 2002. *Publishing and Medicine in Early Modern England*. Rochester/New York: University of Rochester Press.

Gouge, William. 1622. *Of Domesticall Duties: Eight Treatises*. London: John Haviland for William Bladen.

Jajdelska, Elspeth. 2016. *Speech, Print and Decorum in Britain, 1600—1750: Studies in Social Rank and Communication*. Abingdon: Routledge.

Law, Thomas. 1657. *Naturall Experiments or Physick for the Poor*. London: For Edward Farnham.

Leong, Elaine, and Sara Pennell. 2007. Recipe Collections and the Currency of Medical Knowledge in the Early Modern "Medical Marketplace". In *Medicine and the Market in England and Its Colonies, c. 1450–c. 1850*, ed. Mark S.R. Jenner and Patrick Wallis, 133–152. Basingstoke: Palgrave Macmillan.

Mackenzie, George. 1663. *Religio Stoici*. London: For George Sawbridge.

Marsh, Christopher. 2010. *Music and Society in Early Modern England*. Cambridge: Cambridge University Press.

Mortimer, Ian. 2007. The Rural Medical Market Place in Southern England c. 1570–1720. In *Medicine and the Market in England and Its Colonies, c. 1450–c. 1850*, ed. Mark S.R. Jenner and Patrick Wallis, 69–87. Basingstoke: Palgrave Macmillan.

Pepys, Samuel. *The Diary of Samuel Pepys*, eds. Robert C. Latham and William Matthews, 11 volumes (London: 1970–1983), vol. VIII.

Richards, Jennifer. 2015. Reading and Hearing *The Womans Booke* in Early Modern England. *Bulletin of the History of Medicine* 89 (3): 435–436.

Wallis, Patrick. 2007. Competition and Cooperation in the Early Modern Medical Economy. In *Medicine and the Market in England and Its Colonies, c. 1450–c. 1850*, ed. Mark S.R. Jenner and Patrick Wallis, 247–268. Basingstoke: Palgrave Macmillan.

Withey, Alun. 2011. *Physick and the Family: Health, Medicine and Care in Wales, 1600–1750*. Manchester: Manchester University Press.

CHAPTER 8

Archives, Paratexts and Life Writing in the First World War

Hannah C. Tweed

The chapters in this collection have demonstrated a range of ways the phrase 'medical paratext' can be conceptualised, and particularly the inter-actions between medical practice, medical texts, and their writers and readers. Focusing on the diaries of Canadian nurse-writers in the First World War (particularly the work of nursing sister Clare Gass and VAD Alice Lighthall), this chapter proposes that paratext can demonstrate the contemporary archiving and historiography of the authors' experience, and support their claims to authoritative writing—as military historians, as medical practitioners, and as women operating within male-dominated environments.

Nursing already occupied a contested space in war narratives; Margaret H. Darrow describes the VAD as 'the best example of the pervasive unease with any connection between women and war' (Darrow 1996: 82), and many French nurse-writers' accounts of the war disappeared after limited print runs (e.g. Noëlle Roger's *Les carnets d'une infirmière* (1915))—particularly if the authors focused on their experiences over those of the soldiers they nursed. Darrow suggests that the French

H. C. Tweed (✉)
University of York, York, UK

© The Author(s) 2018
H. C. Tweed, D. G. Scott (eds.), *Medical Paratexts from Medieval
to Modern*, Palgrave Studies in Literature, Science and Medicine,
https://doi.org/10.1007/978-3-319-73426-2_8

volunteer nurses of the First World War had 'the best chance to create a story of [French] women's war experience, [and] the fact that no such story entered the culture is significant' (Darrow 1996: 84). While the nurse-writers discussed here are French Canadian, Darrow's comments on dominant, male-centric narratives during and in the immediate aftermath of the war provide a potential explanation for Matron-in-Chief Margaret Macdonald's inability to find a publisher for her history of the Canadian Nursing Services—with contributions from a range of nurses' diaries, memoirs, and recollections of the First World War (Mann, 'Introduction', xxxvi). Meanwhile Andrew Macphail's official history of the Canadian medical services was published in 1925, but made little mention of nursing or VADs. Similarly, an officially-sanctioned volume commemorating the work of the No. 3 Hôpital Général was published in 1928, edited and compiled by R. C. Fetherstonehaugh, with a foreword from 'His Royal Highness the Duke of Connaught, K.G., K.T., K.P., G.C.B., G.C.S.I., G.C.M.G., G.C.I.E., G.C.V.O., G.B.E.' (frontispiece), and supported by the Department for National Defence and the Medical Faculty of McGill University (vii, xi). The text features two images of nurses (out of thirty-eight), and no single chapter is dedicated to the work of nursing sisters or VADs.

There are obvious exceptions to this neglect of nurses' writing—Mary Borden's *The Forbidden Zone* (1929), Ellen La Motte's *The Backwash of War: The Human Wreckage of the Battlefield as Witnessed by an American Hospital Nurse* (1916; repr. 1934), and Vera Brittain's *Testament of Youth* (1933), to name but a few. However, firstly, the majority of these texts were published—or, in La Motte's case, reprinted—in the second wave of war memoirs and writing that explicitly criticised the narratives of glorious war. Secondly, Borden, La Motte and Brittain self-identified as writers before the war, and both Borden and La Motte socialised with modernist writers and society; according to her letters, La Motte 'met a lot of interesting people through the Steins—Gertrude Stein has had me to lunch, and dinner and tea, and has given me her books, two of them' (31 October 1913, The Ellen N. La Motte Collection).[1] Finally, a sizeable majority of the nursing memoirs, diaries and biofictions published in the aftermath of the First World War were written by the women who worked together in Mary Borden's self-funded L'Hôpital Chirurgical Mobile No. 1—a field hospital that Christine E. Hallett refers to as 'not only a centre of healing, but also [...] a cauldron of literary creativity', particularly for nurse-writers (xi).

This latter factor should not be lightly dismissed: in contrast, Clare Gass, the best-known Canadian nurse-writer, is remembered not as a diarist, but for having recorded the first copy of John McCrae's 'In Flanders Fields' (1915), several weeks before he submitted it for publication. Gass's diaries, spanning work in No. 3 Hôpital Général and No. 2 Canadian Casualty Clearing Station from 1914 to 1918, were only published in 2000, edited by Susan Mann. Gass's war diaries, as with many writings of nursing sisters, detail a combination of the everyday, the personal, military history, and a detailed record of medical treatments (numbers of patients, types of injuries or illness, attempted treatments).

In contrast, Alice Lighthall's diaries and collected writings have yet to be published; an exhibition by McGill University Rare Books and Special Collections, 'The Lighthalls: A McGill Family at War' (26 Feb.–15 June 2015) was the first public display of Alice Lighthall's papers. While the archival status of these texts does not lessen their significance, it does highlight the deliberate nature of Lighthall's use of paratext: the loose-leaf insertions to her diaries were meticulously dated and identified (in some cases several years after the end of the war), and explicitly recorded as part of her military and medical service. For example, one of Lighthall's poems, 'I Found Him in the Forest', inserted into the diary entry for its date of composition (29 September 1918), is inscribed on the back with 'Written while on service at No. 5 Hospital, B.E.F., Rouen, France. Alice Lighthall, V.A.D.'. These statements of authorship indicate both Lighthall's military position (part of the British Expeditionary Force in Rouen) and her medical responsibilities (as a serving member of the Volunteer Aid Detachment). Lighthall uses marginalia and peritextual insertions throughout her papers to add authority and accuracy to her observations and writing, and adds clarifying notes on military and medical acronyms, in a manner that indicates an intended audience—even if the papers were not published in her lifetime.

Writing about the professionalisation of historians at the turn of the twentieth century, Elise Garritzen comments on the significance of paratexts for female scholars: how, to signal their professional status, 'women used title pages to demonstrate their qualifications, either by pointing out previous studies they had written or, as Liisi Karttunen once did, inserting their academic degree on a title page' (Garritzen 2012: 413). While Garritzen's analysis of gendered paratexts does not engage with medical practice, both academia and medicine were similarly limited in terms of female access in the early twentieth century—and the dominance of male accounts of the First World War in the period immediately after its conclusion. Despite the lack of public readership for Lighthall's papers, as archival holdings there remains evidence of an intended audience and attempts at the signalling of knowledge and authority discussed by Garritzen. These signals are primarily evident in Lighthall's self-archiving of the papers, both as she wrote within the five-year diary format, and the inserted notes and page-markers listing key battles and significant points in the development of the war. Significantly, in terms of paratext as indicating authorial historiography, Lighthall's diaries were compiled using a 'Walker's "Year by Year" Book', that encouraged the author to write entries for up to five years on each dated page, and compare their present experiences with those of the past (see Fig. 8.1). Lighthall used this format throughout her war service (and earlier, in her teenage diaries), and also preserved a range of ephemera as loose-leaf additions to the diary. I refer to these collected writings and collection as 'the Lighthall papers' and 'Lighthall's diaries' throughout this chapter, considering all of the material Lighthall included as part of the wider text. I argue that this marginalia and ephemera constitutes—and should be viewed as—a self-conscious archiving of Alice Lighthall's war experience, with paratexts employed as a means of demonstrating the different strands of Lighthall's experience: military, medical, and gendered.

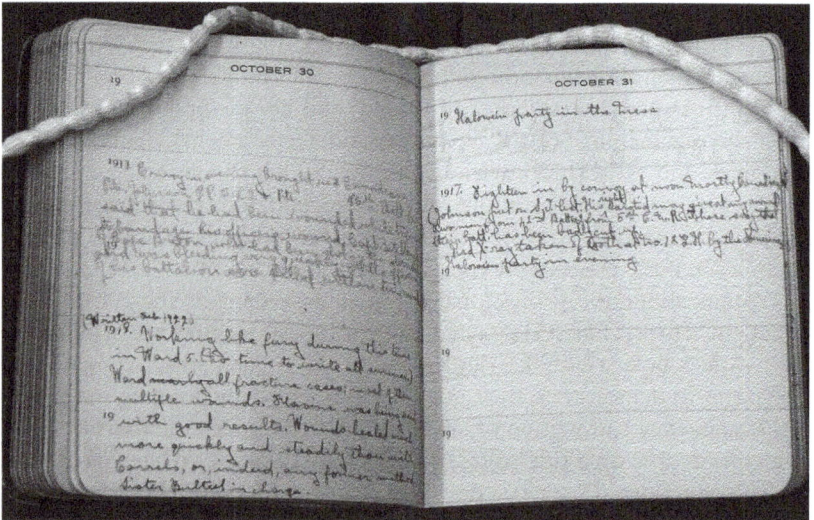

Fig. 8.1 Lighthall Papers, October 30th–31st

Alice Lighthall joined the Volunteer Aid Detachment (VAD) in 1916, from Montreal, Canada, and served as a VAD nurse at the No. 5 Royal Army Medical Corps (RAMC) General Hospital in Rouen from 1916 to 1918. Prior to signing up, Lighthall was a 'partial student' of the Arts as part of the women's programme at McGill University (Quiney 2017: 89), and a member of the Montreal Junior League, a philanthropic organisation for young women (Chenier 2009: 693). As a wealthy upper-class VAD Lighthall was simultaneously of higher social standing than many nursing sisters, and of low medical and military rank—especially in comparison to university-trained nurses like Clare Gass, who signed up for war service via the Canadian Army Medical Corps (CAMC) and in recognition of the increasing professionalisation of nursing were given officer status as lieutenants. As a VAD, Lighthall's position was particularly liminal in terms of the lines between the professionalisation of nursing and the social acceptability of upper-class volunteering. Many medical professionals, including nursing sisters, viewed VADs (who had minimal medical experience) as potentially damaging to the reputation of the nursing profession. Dr. C. K. Clarke, the director of the Canadian National Committee for Mental Hygiene, argued again the inclusion of VADs in hospitals, as 'the unqualified woman is a nuisance' (Quiney 2017: 39), while Matron-in-Chief Margaret Macdonald refused to accept VADs as part of CAMC military nursing at the Front (Quiney 2017: 8). Lighthall's diaries reflect these concerns over medical experience, class, gender, and her status as contested historical observer.

Both the content and the form of Lighthall's diaries are significant. The five-year diary is a development of the nineteenth-century pocket diary, which Sandro Jung describes as essentially conservative in form—compact, with illustrated pages, but a form that 'did not offer women the opportunity to promote their own work until the mid-nineteenth century', even though pocket diaries were marketed explicitly at women (Jung 2012: 29). One might, in the *Le Souvenir, or Pocket Table* (1842), a popular pocket diary, have 1–15 August on one page, with an illustration of Borrowdale, in the Lake District, at the top of the page (Jung 2012: 38). Women, presumably, did not need more than one line to write about their daily activities and experiences—just a few key words. Both Gass's and Lighthall's multi-year diaries feature entries where the page became a palimpsest when space was at a premium, as well as short epigrammatic entries; writing up as well as across the page and the central divide, in an attempt to convey their experiences beyond the space given to them (e.g.

Lighthall, 14 August 1918, or Lighthall, 11–12 November 1918). The pages also swell with inserted ephemera—handwritten music scores from Herbert Howells' 'Magnificat and Nunc Dimittis' in E (20 July 1918), Lighthall's poetry drafts, bookmarks indicating major battles or medical events, and newspaper cuttings memorialising colleagues. Nancy Martin describes the diary as 'defy[ing] the traditional structural forms of narrative. Focusing generally on the immediate present, it is serial, open-ended, often repetitive and contradictory' (Martin 2015: 1248). Martin details the attention to preserving minutiae in First World War diaries; in particular, 'photographs, a lock of child's hair, flowers from the family garden [...] sketches of weapons, open fields, destroyed buildings, and trenches' (Martin 2015: 1248). Items such as those described above fall squarely into the category of paratext, particularly if they have been preserved within archival contexts, and they are found in both Gass's and Lighthall's work (Gass preserving photographs—taken illegally, with her own camera—newspaper articles, and botanical cuttings in her diary).

First World War diaries have been commonly presented as 'a private, self-reflective form', that demonstrates the actions and emotional states of individual men and women (Martin 2015: 1245). Martin argues that diaries provided a space in which the soldier-diarist could 'reevaluate his newly militarised identity', even as the government 'worked to recode (and control) the wartime behaviour of its citizens' (Martin 2015: 1247)—whether by drawing on military metaphors and classical allusions, or by presenting fragments of narrative, as with Mary Borden's *The Forbidden Zone*, her 'collection of fragments' that reflected 'a great confusion' (Borden 2008: 8). If one extends Martin's arguments across to the multi-year diaries used by Lighthall and Gass, I would suggest that as well as negotiating their identities as women and medical professionals (with varying degrees of training and authority) operating at the Front, Lighthall and her contemporaries—particularly those who, like Lighthall, joined up part-way through the war—also demonstrate a negotiation of their self-conscious status as historians, as well as witnesses and participants. More than simply representing self-reflective writing, these diaries indicate an awareness of the significance of the authors' status as both witnesses and medical practitioners.

While I describe Lighthall and her contemporaries as participants, both in terms of medical service and writing, these diaries are also curiously distant from more confessional forms of autobiographical writing. Leigh Gilmore, in *The Limits of Autobiography: Trauma and*

Testimony (2001), suggests that individuals who are narrating trauma frequently distance themselves from autobiography. She argues:

> Because testimonial projects require subjects to confess, to bear witness, to make public and shareable a private and intolerable pain, they enter into a legalistic framework in which their efforts can move quickly beyond their interpretation and control, become exposed as ambiguous, and therefore subject to judgements about their veracity and word […] although those who can tell their stories benefit from the therapeutic balm of words, the path to this achievement is strewn with obstacles. To navigate it, some writers move away from recognizably autobiographical forms even as they engage autobiography's central questions. (Gilmore 2001: 7)

This description holds true for both Gass and Lighthall. Instead of metaphorical or emotive language, emotional affect is conveyed by both diarists via juxtaposition of subject matter—the everyday and quotidian alongside the horrific and bloody, with no explanatory comments linking the two. For instance, Gass wrote on 27 October 1915 of overseeing blood transfusions, and on 30 October 1915 of receiving 'a big convoy of patients […]. Capt Burgess opened up a new line of Hubert Neilson tents to serve for his ward for the present. Miss Eastwood & I went there this morning […] We are still able to get lab roses for our dining tables. It has been a wonderful summer of Poppies & Roses' (75–76). Given that Gass then transcribed McCrae's 'In Flanders Fields' (with some differences in punctuation and indentation) immediately after this entry, there seems a deliberate irony to the juxtaposition of the medical and the pastoral. Similarly, on 25 March 1918, Lighthall wrote: 'Went to dell in the forest for flowers, found wood anemones and primroses. Got back to find the ward turned surgical. First convoy of rush in. Dressing till 10.30pm.' The frequently jarring juxtaposition of subject matter in these diaries (part personal travelogue, part medical and military history) is even more pronounced when entries for multiple years are compared simultaneously, as the war developed.

This use of form to provide reflection and highlight significance was part of the intention of the publishers of the multi-year diary (albeit probably with more domestic spheres in mind than war hospitals). 'Walker's "Year By Year" Books', according to their prefatory material, 'set forth an altogether new and novel idea and one that is eminently useful' (iv). The preface continues as follows:

It means little and much: little recording and much satisfaction.

Many have neither the time nor the inclination to keep a full Diary. But out of the multitude of matters that crowd the experience of each day, there is always something that intelligent people desire not to let slip, but seek to hold 'to awaken memory' in days to come. What a record of experiments such a book may be made! – things accomplished, things attempted, successes, failures, joys and sorrows.

For just such a record this book is designed, and it is so planned that each day may be compared with the corresponding one in any year for five years. Five years hence, if you have kept your record faithfully, you will undoubtedly be deeply interested and delighted to open the book anywhere and see how wonderfully the web of your life has been woven. It will be a complete story, for you will come to rejoice in briefly expressing what you wish to record, and treasure it in proportion to its brevity, easily recognising and sifting the important from the comparatively trivial. (iv)

Furthermore, the format also provides the reader with a greater variety of reading options than is typical. The Walker prefatory material continues:

To illustrate how it should be used. You may commence any day of the year, but suppose that you begin January 1; under that day, in the first space, add the proper figure for the year. On the next day, January 2, do likewise, and so on throughout the year. When the year is ended begin again under January 1 for the second year, add the appropriate figure in each of the second spaces, and so right on through the remaining years. (iv)

As a result of this system, Alice Lighthall's war diary 'begins' (for the reader) on 3 January 1919, with what are chronologically her last two entries. Paratextually, the first entry details Lighthall meeting a friend at Bridgwater railway station, Somerset, at the end of the war, before sailing home to Canada. To begin a war diary with the author's discharge from military duty, holidaying in England, is atypical (although there is a certain symmetry with Lighthall's chronological first entries, when she visited a series of friends in Montreal and London before beginning her training as a VAD). The abrupt contrast with the entry for 11 January 1917 (which mentions 'rumours of zepp. [zeppelin] raid', is clear. Such a format also encourages the reader to approach Lighthall's text repeatedly, and from multiple directions. Should one read each entry chronologically, by calendar composition? Does one start on 1 January, and read each date page, comparing multiple years at a time? Similarly, how should the reader respond to authorial restructuring of diaries? Clare Gass thriftily re-used her 1915 diary to incorporate her observations during 1916, altering the days as 1916 was a leap year (Mann 2000: xli). These are also questions for the publisher and editor: Susan Mann, in editing *The War Diary of Clare Gass* (2000), explicitly states that she chose to follow a chronological, entry-by-entry system in publishing Gass's diaries—cutting out the multi-year structure entirely. Mann summarises this editorial decision as follows: 'I have chosen not to impose [Gass''s] practical but idiosyncratic and occasionally confusing format on the readers of this publication; I have also checked her sign-posts to ensure that the two years are distinct' (Mann 2000: xli). While this choice makes for a simpler and more coherent reading experience, I suggest that nonetheless it alters a key element of the paratext, and thus the content of Gass's diaries.

I have described both Gass's and Lighthall's diaries as concerned with military, medical, and personal experience during their respective service in France. In Lighthall's papers, examples of this combined content includes her early account, on 19 February 1917, of the funerals of a VAD from No. 9 General Hospital and a nursing sister from No. 11 General Hospital:

> Funeral of V.A.D. […] and an Australian Sister […] both of whom had died of spinal meningitis. Buried in our part of the cemetery here. Saw graves of several Canadians there, among them Major Moss, 3ʳᵈ Battalion. Mostly from Langemarck [Third Battle of Ypres], or the push last autumn.
>
> Rush began today. Convoys in continually from now on. Ports closed, so none out. Most cases surgical. Five tressles [sic] put down in each ward.

The references to specific diagnoses and cause of death—spinal meningitis—are characteristic, as is the reference to 'tressles', or how overcrowded Lighthall's ward was, with additional stretchers placed on trestles in the aisles in lieu of beds. A subsequent entry on 21 February 1917 refers to how Lighthall 'was to have had half-day, but "convoy" blew during second lunch hour, so that was off. Seven patients received in 15. Six extra tressles [sic] ordered raising our capacity to 42. All of patients and two admissions put on tressles.' This pattern of brief narratives of limited medical supply (precisely detailed) and human injury continues in Lighthall's entries for March and April 1917, tracking the medical development of the war, with a 'suspected case of Scarlet Fever in 15', where the soldier was 'isolated in side-ward, and taken away […] in evening', before the ward was placed under 10 days' quarantine (3 March 1917). The double-spread for 11 and 12 April 1917 details 'surgical convoys nearly all day. 17 admitted to 15. Heavy cases and light mixed. [...] 15 Ward completely filled. Operations and dressings all day.' In terms of the history of disease and infection in the First World War, cerebro-spinal meningitis and sepsis were common causes of death (with epidemics of meningitis across Europe and North America during the First World War), and scarlet fever was prevalent. Most significantly, in terms of medical and world history, Lighthall describes the outbreak of Spanish Flu, which would kill over 20 million people worldwide (Weitzman 2001; Ash 2014):

> Largest death-toll tonight that the hospital has ever had. More convoys of the new 'flu', almost like Pneumonic Plague. Starts with head symptoms [...] then develops into violent pneumonia, with infamous and wild delirium. Temps turn 104 and 105 for a day or two, then death follows. 3 Ward is to be taken over for it next. All the medical side is crowded with it, also Wards 1 and 2. Gas marqhees [sic] busy. (Smith moved there.) All wounds where it is are isolated, nurses wear gauze masks.
>
> Unofficial news tonight that Austria has surrendered unconditionally. Can it be true? No more drafts from here being sent at the time. (1 November 1918)

The description of a flu 'almost like Pneumonic Plague', and the increasingly detailed accounts of symptoms and medical responsibility are particularly jarring given the earlier entry, from 1 November 1916: 'Went to the cemetery, and put fresh flowers on the graves of all the Canadians I could find (Major Moss [...] the only officer among them.) All the French graves wonderfully decorated, and people making pilgrimages to them.' In the 1916 entries, Lighthall had the time and inclination to visit the graves of her countrymen; in 1918, not only had she had 'no time to write all summer' (20 October 1918), but her entries are markedly less likely to mention the class of her patients, and the proportion of officers to privates and NCOs in the ward. Instead, Lighthall's comments are focused on medical precautions, and brief speculations on the development of the wider war effort.

A similarly contrasting pattern is seen in Lighthall's diary entries for late March. In the same double-paged spread, Lighthall refers to:

> [The] ward filled to overflowing. Extra bed put in, five stretchers put down centre of ward. Dressing all day, up till 9pm. More refugee sisters arrived. Sister Blades told me it was wise to be ready to pack up at short notice, because we might possibly have to evacuate if Amiens falls. (28 March 1918)

Meanwhile, on 29 March 1917:

> Wild discussion broke out in the ward among English, Scotch and Australian men, about who were the bravest soldiers in France. They gave that honour to the Canadians, as having held the hottest front of the British line longer than anyone else. McGeachan told me about the 2nd battle of Ypres, and of how the Canadians lost their guns, and then won them back <u>again</u>, and immediately turned them on the Germans. He said that was the only time the Germans had used a gas […], as it proved too dangerous to themselves.

The uncomplicated patriotic fervour of Lighthall's entries for 1916 and early 1917 do not feature in her chronologically later entries, as her responsibilities and the death toll rose. While a shift in perspective is hardly uncommon in war diaries and accounts, this contrast is rendered particularly striking and obvious by the multi-year format of Lighthall's diaries. The relative optimism of the chronologically earlier entries, and the narrative significance of the multi-year form are most obvious in the short double-spread of 15 and 16 November:

> 15 Nov 1918: Put in application to have my contract cancelled.
>   16 Nov 1916: Signed on for the next six months, or the duration of the war, whichever terminates first.

Writing about VAD narratives and personal writings, Linda J. Quiney comments how in addition to Governmental censorship, 'hospital convention imposed its own censorship, well understood by the nurses and impressed upon the VADs, that they were not to discuss the private details of the patients or procedures' (Quiney 2017: 9). Quiney summarises many of these self-censored narratives as 'a cheery letter home to family or friends describing the delights of an afternoon outing with colleagues or the beauty of the French countryside, with no mention of the stresses or tensions of the hospital ward' (Quiney 2017: 9). While some of Gass's and Lighthall's writing does conform to this 'cheery' description, there are also longer narrative pieces and poetic extracts to both texts—and their self-editing is historiographically significant. Lighthall's main entry for 14 August 1918 reads as follows:

> Air raid last night shortly after eleven. I hadn't gone to sleep, though Bundett had. Both had to get up and dress and go out to our trenches. Could hear machines plainly for a long time right overhead. Barrage was

lively, especially machine guns just across the road. Dropped an incendiary bomb which lit up the whole place. (Heard today that it hit the Rue Verte station.) Trenches very uncomfortable. Word came after everything had been quiet about 15 minutes to get out of them. Went up, only to be sent back by Matron. Then someone fainted further along, so in about twenty minutes we were bellowed to come out to go to our rooms. Told to go to medical wards if another alarm.

Second alarm about half-past twelve. Machines close at hand again. Got up and went to Ward 13. (Bundett wouldn't come then.) Found Bundett with a flash-light, very nervous, patients sleeping quietly although the noise outside was terrific. Soaked a blanket under kitchen tap in case of fire. Both the day orderlies were there [...] We all went to sit in the bunk until the noise stopped. Nothing seemed to be hit near us. Heard Capt. Lang ordering his patients to "stay where they were until the 'All Clear' blown". They seemed to be in the trenches outside 11 Ward. Went home about one mostly dead with sleep. Found everyone else had come back earlier.

Third alarm blew about half past two I think. Dressed again, but didn't go out. Machines further away. Barrage not heavy. "All Clear" blew in about half an hour.

[Written up the spine:] Heard today that the [illegible] Hospital was hit, also the theatre in the Rue de la République. Bombs dropped in R. de la R. and Rue Verte besides.

In addition to the framing of these air raids as a fleshed-out first-person narrative, Lighthall's focus on patient outcomes as well as personal experience is clear. Elsewhere, Lighthall mentions a colleague's account of air raids at the No. 8 Scottish General Hospital, where 'patients, barely able to walk themselves, helped to get the wounded out, and carry them to cellars', and in the aftermath of three weeks of bombing 'some of the VADs were shell-shocked, and [were] sent home on leave' (28 October 1917). These air raids were evidently an episode that Lighthall revisited—firstly, in a pencilled note in the margins of her entry for 14 August 1918, that 'Think this was the Rive Gauche [Left Bank] station, which was wrecked', with an arrow pointing to the hospital named in the main entry. Additionally, Lighthall inserted a bookmark into page spread for 15 and 16 August, marked '1918 Air Raids' (see Fig. 8.2). Such additions, far from damaging the original text, demonstrate a self-conscious desire to present a historically accurate personal account. Evidence of careful and deliberate archiving is evident elsewhere in Lighthall's papers: in the entry

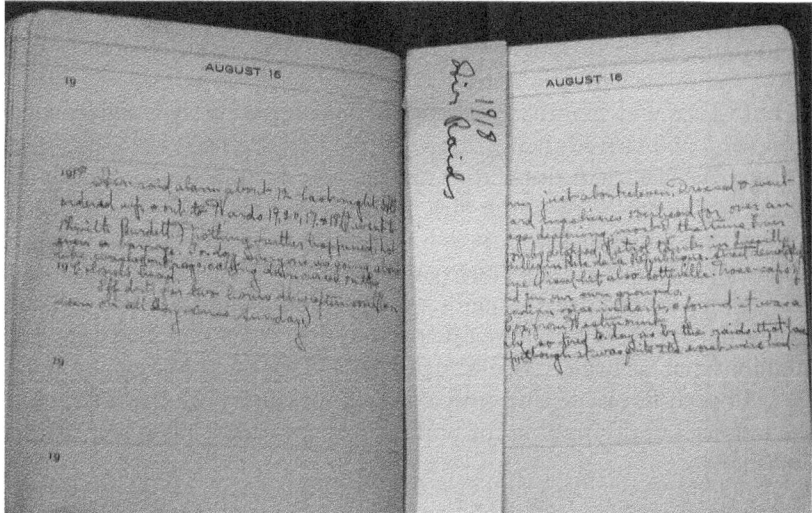

Fig. 8.2 Lighthall Papers, August 15th–16th

for 8 April 1917, Lighthall wrote 'See May 8. Mistake in dates'—presumably with an audience other than herself in mind, and a desire for a coherent narrative.

These bookmarks are the final point of paratextual control in Lighthall's diaries—seen in the entries for 14–16 August 1918, but also throughout the text. Functioning as historical and personal chapter divisions, they include titles such as '1918 Amiens Push' (25 March), '1917 Passchendaele/1918 Ward 5' (6 August), '1916 Departure' (17 September), '1916 Arrival in France' (18 October), and 'Armistice Day' (11 November). Gérard Genette categorises this kind of material as '*intimate epitext*'—or 'any message bearing directly or indirectly on an author's own past, present, or future work which the author addresses to himself, with or without the intention of publishing it later' (Genette 1997: 387). I suggest that while these bookmarks—and, indeed, all of the paratextual material comprising Lighthall's self-archiving and editing of her papers—can be considered in these terms, there is something more political to the act of self-archiving, for both Gass and Lighthall, and their contemporary nurse-writers. In clearly presenting their work as the account of nursing practitioners, observing and participating in military and medical history—both masculine-dominated zones—their careful attention to paratextual detail indicates authorial recognition and control of the significance of their narratives.

Both Gass and Lighthall operated within a context that undervalued their work and viewed their writing and medical practice as socially problematic. As such, it is not surprising that few nursing diaries were published in the aftermath of the First World War. However, the continued archival and publishing neglect of war nursing diaries also raises the question of whether we are still ill-prepared to publish the paratextual features that complicate these texts. Digitisation offers one solution to the paratextual problems of publishing war diaries and their associated insertions and additions; but, as Robert MacLean discusses in "Medical Marginalia in the Early Printed Books of University of Glasgow Library" in this collection, the politics of prioritisation for both cataloguing and digitisation remain an issue.

NOTE

1. With thanks to Alice Kelly for directing me to these letters in the La Motte Collection.

BIBLIOGRAPHY

Ash, Caroline. 2014. What Made Spanish Flu So Deadly. *Science* 344 (1688): 1129.

Borden, Mary. 2008 (orig. 1929). *The Forbidden Zone*. London: Hesperus Press

Brittain, Vera. 1933. *Testament of Youth*. London: Victor Gollancz.

Chenier, Elise. 2009. Class, Gender, and the Social Standard: The Montreal Junior League, 1912–1939. *The Canadian Historical Review* 90 (4): 671–710.

Darrow, Margaret H. 1996. French Volunteer Nursing and the Myth of War Experience in World War I. *The American Historical Review* 101 (1): 80–106.

Fetherstonehaugh, R.C. 1928. *No. 3 Canadian General Hospital (McGill): 1914–1919*. Montreal: The Gazette Printing Company.

Garritzen, Elise. 2012. Paratexts and Footnotes in Historical Narrative: Henry Biaudet and Scholarly and Nationalistic Ambitions of Historical Research 1902–1915. *Scandinavian Journal of History* 37 (4): 407–429.

Gass, Clare. 2000. In *The War Diary of Clare Gass: 1915–1918*, ed. Susan Mann. Montreal: McGill-Queen's University Press.

Genette, Gérard. 1987; trans. 1997. *Paratexts: Thresholds of Interpretation*. Cambridge: Cambridge University Press.

Gilmore, Leigh. 2001. *The Limits of Autobiography: Trauma and Testimony.* Ithaca: Cornell University Press.

Hallett, Christine E. 2016. *Nurse Writers of the Great War.* Manchester: Manchester University Press.

Jung, Sandro. 2012. The Illustrated Pocket Diary: Generic Continuity and Innovation, 1920–40. *Victorian Periodicals Review* 45 (1): 23–48.

La Motte, Ellen. 1913. *Letter, 'Paris, 166, Bld. Montparnasse, 31 October, 1913', Ellen N. La Motte Collection.* Baltimore: The Alan Mason Chesney Medical Archives of The John Hopkins Medical Institutions.

———. 1916; repr. 1934. *The Backwash of War: The Human Wreckage of the Battlefield as Witnessed by an American Hospital Nurse.* New York/London: G. P. Putnam's Sons.

Lighthall, Alice. 1916–1919. *Lighthall Papers.* Montreal: McGill University Rare Books and Special Collections.

Macphail, Andrew. 1925. *Official History of the Canadian Forces in the Great War, 1914–1919: The Medical Services.* Ottawa: F. A. Acland.

Mann, Susan. 2000. Introduction. In *The War Diary of Clare Gass: 1915–1918*, ed. Susan Mann, xx–xi. Montreal: McGill-Queen's University Press.

Martin, Nancy. 2015. "And All Because It Is War!": First World War Diaries, Authenticity and Combatant Identities. *Textual Practice* 29 (7): 1245–1263.

Quiney, Linda J. 2017. *This Small Army of Women: Canadian Volunteer Nurses and the First World War.* Vancouver: UBC Press.

Roger, Noëlle. 1915. *Les carnets d'une infirmière.* Paris: Attinger Frères.

Weitzman, Jonathan. 2001. Spanish Flu. *Genome Biology* 2. https://doi.org. ezproxy.lib.gla.ac.uk/10.1186/gb-spotlight-20010910-01. Accessed 22 Sept 2017.

"Nonsense Rides Piggyback on Sensible Things": The Past, Present, and Future of Graphology

Deborah Ellen Thorpe

"Nonsense rides piggyback on sensible things", declares professional sceptic and questioned-document analyst Joe Nickell in his chapter of *The Write Stuff* (1992: 15).[1] Nickell refers to graphology's influence despite its failure to pass scientific validity tests. According to the British Institute of Graphologists (2015), graphology is the study of the "movement, spacing, and form" of handwriting. Handwriting presents both individual difference and internal variability, and practising graphologists use these variations to support statements about an individual's personality, mental health, and emotional state. Nickell's scathing metaphor expresses his qualms with the way in which graphologists exploit the wealth of information contained in handwriting to justify dubious assertions.

The original version of this chapter was revised. An correction to this chapter can be found at https://doi.org/10.1007/978-3-319-73426-2_11

D. E. Thorpe (✉)
Trinity Long Room Hub Arts and Humanities Research Institute,
Trinity College Dublin, Dublin, Ireland

© The Author(s) 2018
H. C. Tweed, D. G. Scott (eds.), *Medical Paratexts from Medieval to Modern*, Palgrave Studies in Literature, Science and Medicine, https://doi.org/10.1007/978-3-319-73426-2_9

Several well-respected applications of handwriting analysis scrutinize variations between, and within, scripts. For instance, questioned-document analysis in forensics examines handwriting features to detect forged signatures, identifies alterations in documents, and recognizes the author of anonymous script (Nickell 1992: 43). Palaeographers—researchers of the forms and processes of historical writing—might base an identification of an anonymous scribe on the most unconscious features of writing. Good practitioners in these fields are willing to admit the limitations of their methodologies without detracting from their overall validity. For instance, a palaeographer might advise that scribal hands vary on a day-to-day, or even an hour-to-hour, basis, which makes definitive identifications difficult.

In contrast, graphology is unique in its claims to be able to recover "the elusive quality of a writer's personality" from handwriting alone, without—as Nickell argues—standing up to rigorous academic examination. In addition, graphologists have been reluctant to admit their failures or define and justify their methodologies, likely due to a combination of professional competition, protectiveness of their reputation, awareness of the sceptical viewpoint of academic researchers, and a belief that scientific testing "oversimplifies" their work (Lockowandt 1992: 94). Finally, they neglect what many researchers perceive as one of the most useful applications of handwriting analysis—for the objective assessment of the movement disorders caused by different neurological conditions—choosing instead to focus on specious declarations about an individual's personality or character.

In a digital age, when it is possible to self-publish to diverse audiences through websites and social media, graphologists are finding louder voices online—as are the sceptics. Graphology is no longer restricted to the "occult sections of local libraries" as it was in the 1990s, but is easily discoverable on the internet (Beyerstein and Beyerstein 1992: 15). A search for "#graphology" on Twitter returns copious tweets by and about graphologists, indicating that it is the subject of lively discussion online. The topic of graphology has been investigated by many online magazines and news outlets in recent years. Coverage ranges from light-hearted curiosity, as in *The Guardian*'s piece "Prince Charles letters: What does a graphologist make of them?" (Khaleeli 2015), to ridicule in *The Spectator* blog's response: "Charles's 'spider letters': *The Guardian* falls for the pseudoscience of graphology" (Thompson 2015).

The handwritten letters of infamous criminals are particularly enticing to journalists, since they can be analysed retrospectively for clues that the crime was inevitable. Thus, graphologists are routinely ushered into the limelight to speculate about the mental health and criminal proclivities of certain writers. For instance, in 2016, the *Daily Mail* interviewed graphologist Brigitte Applegarth about letters written by the murderer of Labour MP Jo Cox: "[his] handwriting shows he has an inferiority complex and wishes to punish those who do not agree with him, according to an expert" (Greenwood and Sinmaz 2016). Whether positive or negative, these articles about graphology have increased its reach to public audiences. With this in consideration, I investigate whether graphology—the interpretation of handwriting as a paratext containing clues about the writer's personality—has gained any perceived legitimacy from its new visibility.

Touching on medicine specifically, I first ask: is it possible to metaphorically "dissect" the page of handwritten texts, to scrutinize writing as a "medical paratext" rich in information about the writer's state of health? Medical research suggests that it is: neurologists and psychiatrists are advancing investigations into the relationship between handwriting performance and brain activity. Studies have used writing as a tool for medical diagnosis, finding certain handwriting features to be reliable biomarkers for conditions such as Parkinson's disease (Rosenblum et al. 2013a). The effect of age-related deterioration on writing processes is also well documented, with handwriting changes being symptomatic of deterioration in the frontal and prefrontal cortex (Rosenblum et al. 2013b). Thus, there are uncomfortable synergies between graphology and medical handwriting assessment, both of which make connections between human differences and individual difference in handwriting. However, though the connection between medical pathology and handwriting performance is widely accepted, the relationship between psychological traits and states and writing is controversial. Despite this, even graphology's most vocal sceptics acknowledge that it is possible to learn *something* about a person, aside from their physical health, from their handwriting: clues about their gender, nationality, and profession, for instance (Nickell 1992: 46).

It is unsurprising, given its reputation as pseudoscience, that medical practitioners endeavour to sever any links with graphology. However, more surprising perhaps is that graphologists themselves have distanced themselves from medical diagnosis. For instance, graphologist James Crumbaugh (1992) has stated that "neither mental nor physical disease can be diagnosed by Graphoanalysis, but handwriting often provides information that helps the physician make a better estimate of the cause of the symptoms" (Crumbaugh 1992: 177). Taking an unusually cautious stance for a graphologist, Crumbaugh argues that an analysis of handwriting distortions can contribute to, but not determine, a medical diagnosis. This caution is uncharacteristic of a discipline that otherwise makes sweeping statements based upon handwriting features alone: for example, "'[Prince Charles' writing is] a nice clear hand, going logically forward,' [Graphologist Elaine Quigley says], which shows 'he is very much a man who likes to do things his own way'" (Khaleeli 2015).

It is possible that Crumbaugh's position is attributable to fear of his largely intuitive practice being undermined by empirical medical evidence. If so, this highlights a gap in the existing research: we need to know more about the *nature* of the connection between physical and mental states and handwriting. In fact, the boundary between "physical" and "mental" states, itself, is not clear, since there is currently active discussion among neurologists about what comprises a "psychiatric" disorder versus a "neurological" disorder—with some that have been previously labelled "disorders of the mind" being recently re-classified as having possible organic causation (Newby et al. 2017). Thus, the final section of this chapter demonstrates how academics are finally moving to fill this gap, going "back to basics" with their inquiries into individual difference and handwriting features.

This chapter is an updated study of graphology, providing a wider understanding of the concept of the paratext by considering the information captured in handwriting in the context of a digital age. It builds upon Barry and Dale Beyerstein's *The Write Stuff* (1992) to push the evaluation of this controversial, but well-known, discipline into the twenty-first century. There is a tendency to ignore graphology, to assign it to the past as "consigned to the dustbin of history" along with phrenology and astrology (Douglas 2015). However, the synergies and potential overlaps with other types of handwriting analysis should encourage us to make a more constructive criticism of graphology and its impact. Therefore, this chapter outlines the controversies and continuing allure of graphology in the twenty-first century, inspects the tangle between the types of handwriting analysis, and considers whether the idea of a connection between handwriting and subjective aspects of identity (such as psychological character) should be rejected entirely.

GRAPHOLOGY, ITS CONTROVERSIES, AND ALLURE, IN THE TWENTY-FIRST CENTURY

Graphology retained some popular appeal at the close of the last century, despite the lack of experimental support for its principles, and disputes about its validity prompted the publication of *The Write Stuff* in 1992, claiming in its blurb to give "a balanced evaluation" of the practice.

However, the dawn of the twenty-first century brought louder sceptical voices, as regulators realized that graphology was a potential means of unlawful discrimination. The use of graphology in employee selection and monitoring by some companies led the Chartered Institute of Personnel and Development (CIPD) to state that "graphology does not provide a sufficient basis on which to make important decisions about selection, developmental potential, redundancy or aptitude for training" (1998/2001; Bradley 2005b).

In 2005, marketing expert Nigel Bradley expressed his anxiety that this condemnation of graphology, and the embarrassment of companies using it, might lead to its demise—"it runs the risk of becoming a knowledge which is 'lost' knowledge". Worse, he argued that critics had begun to avoid the topic entirely: there had been "an increase in the number of people who do not express an opinion on graphology" (Bradley 2005b). Consequently, students did not choose to study graphology, instead going for "'safer' subjects that are well known to prospective employers". Seeking to regulate graphology and recover its image, Bradley devised a "Code of Conduct for Graphology in Europe" (Bradley 2005a). Amongst his suggestions, which encouraged graphologists to offer an ethical and transparent service, he recommended a disconnect from what he sees as disreputable areas, such as "the occult" ("hypnosis, astrology, mediums, telepathy, and the paranormal"). As a result of graphology's association with the supernatural, he argued, it was considered "a popular toy that has no place in academia or the world of work". In a presentation given to graphologists in 2002, Bradley investigated the relationship between graphology and academia, concluding that "graphology needs to be present in academia ... to develop ... and to be accepted as a legitimate tool".

In preparing *The Write Stuff*, editor Barry Beyerstein undertook a search for such academic representation; he looked for "any independent support for graphological claims in the scientific literature" (Beyerstein and Beyerstein 1992: 15). He was unsuccessful, and instead was "forced to rely on garage sales, New Age booksellers, self-published tracts by graphologists themselves, and the occult sections of local libraries to find the majority of pro-graphology works" (Beyerstein and Beyerstein 1992: 15). If he were undertaking this task today, his search would be faster: the internet proliferates with guides to graphology; institutional pages for graphology organizations; pro-graphology articles (in the popular press and, more rarely, in academic journals); and the professional webpages of practising graphologists.

However, the democratic nature of internet content also amplifies the controversies surrounding graphology. For instance, Wikipedia, being a collaborative encyclopedia that is subject to constant change, creates its own paratexts, by recording alterations to each article in its "article history" (see "How to Read an Article History" 2015). Its "Graphology" article allows unrivalled insight into modern critical responses to the practice, since it registers who has worked on, and who has contended, information in the article (2017, last edit May 6). Treated as a primary text, this Wikipedia article and its paratext offer an opportunity to look over the shoulders of its readers and editors, many of them lay or non-specialist (or, at least they do not claim relevant qualifications) as they engage with the definition of, history of, and potential future of graphology. For instance, user "Tronvillain" (three edits on 12 November 2016), who describes themself as having a bachelor's degree in biochemistry, makes amendments to the article's section on the history of graphology ("User:Tronvillain" 2016). Another non-specialist editor visible in the article's paratext is user "Geeveraune" who provides regular minor edits to the article on graphology in October and November 2016, but otherwise edits articles on geographic information system (GIS) software, historical battles, and places in Ireland ("User: Geeveraune" 2016).

As might be expected, the "Graphology" article is highly unstable—amended regularly by both graphology supporters and sceptics. In the period July to September 2015, it received 27 edits, only six of which were classified "minor". The most frequent reasons for these were disputes over how to evaluate the practice. Most edits propose contrasting critiques of the practice, with graphology supporters and sceptics editing head-to-head. Tagged reasons for edits include "non neutral change", "not interesting biased", and studies being "not recent".

Many of the quickly-reverted edits are non-neutral, unsupported by evidence, and expressed in ungrammatical English. For instance, on 1 June 2015, "Dr Raghvendra Kumar" contributed the following change:

> "'Graphology' is the much more advance science than psychology and psychiatry … based on involuntary expression that is handwriting rather than voluntary expression which is opinion which can not be true because of infinite factors like ego, understanding, motive, circumstances, state of mind and so on." Within just three minutes, the user "McSly" had revised the article, deleting the new content, re-inserting the word "pseudoscientific", and adding the reason "non neutral change". These edits display active policing of the article by numerous individuals, resulting in a consensus on graphology. The article's

current version upholds the sceptical perspective: "Graphology is the analysis of the physical characteristics and patterns of handwriting purporting to be able to identify the writer, indicating psychological state at the time of writing, or evaluating personality characteristics. It is generally considered a pseudoscience."

Though Wikipedia is not intended to be the sole source of reliable information, its dynamic, democratic, and accessible quality makes it a useful gauge of popular opinion—it is a "common resource of human knowledge" ("General Disclaimer" 2015). Internet users seeking an overview of graphology in 2018 will gain an overwhelmingly sceptical perspective from its Wikipedia entry, as its authors parade a series of pejorative adjectives: "worthless", "negative", "illegal", and "vague". However, the persistent allure of graphology should prompt us to ask the question: what *can* we learn about personality from handwriting, if anything? How does writing perform as a paratext, in relation to the meanings and contents of the words and phrases themselves? Is it possible to salvage the foundational principle of the practice of graphology: that there is *some connection* between handwriting features and the more subjective elements of human identity?

"The Careless Flourish": What *Can* We Learn About Individual Difference from Handwriting?

Handwriting has long provoked affective reading. In the Middle Ages, though even intimate letters were often written by scribes, autograph writing was occasionally used as a symbol of affection (Williams 2001: 213). Bernard of Clairvaux closed a letter with the autograph postscript: "I dictated these things but wanted you to recognize my love by a handwriting known to you" (Ganz 1999: 284; Williams 2001: 213). A member of the fifteenth-century Oxfordshire gentry, Elizabeth Stonor, used autograph postscripts—executed with considerable effort in her unpractised hand—to communicate personal news to her husband (Thorpe 2015: 87). For early modern correspondents, autography became more widespread, as a way of ensuring authenticity and representing the true will of the signatory, and so became a symbol of security, intimacy, and trust (Williams 2001: 225). This new prominence of autograph writing enabled writers to capture, and the reader to find, an element of their identity in the handwritten word—something that evades both the writing of an amanuensis and the orally-delivered message of a letter bearer (Williams 2001: 69).

In our current age, where more and more information is recorded digitally, we are nostalgically captivated by handwriting. Though we write by hand increasingly rarely, the handwritten text holds special significance due to the uniqueness of its form. Artist Tim Murray-Browne, discussing his 2016 show "Movement Alphabet", has argued that this is because "handwritten text communicates character and mood"—it is the "residue" of a moving body, and thus the writing enables us to feel a connection with the person who wrote it. He describes an experience that is common to most of us—that of receiving a letter addressed with familiar handwriting: "I receive few handwritten envelopes in the post. I instantly recognize writing on the front for nearly all of them" (Murray-Browne 2016). Implicit in this statement is the emotional reaction provoked by recognition, for example, the joy of laying eyes on the handwriting of a beloved friend. This fascination with the individuality of handwriting has resulted in a proliferation of accounts of the discovery and subsequent curious—and sometimes highly emotional—investigation of handwritten letters and diaries by historical people.

Alexander Masters, in *A Life Discarded* (2016), recounts his compelling story of finding the diaries of an unknown woman, discarded in a skip. His article in *The Guardian* outlines the gradual process of discovery, as the unknown woman's identity becomes clearer, jointly through her handwriting features ("the measured way the writer records the date in blunt, soft pencil"), and the snippets of information and eventually a name—"Laura"—that slowly emerge from the contents of the entries. As Masters learns more about Laura, he connects elements of her psychological state to her handwriting ("Laura's handwriting collapses with her spirit"), and he eventually employs two graphologists. One of these experts, the book reveals, tells him: "the person who has handwriting like this is a complete nutter", an unscientific and unhelpful response.

Despite this graphologist's frivolous inferences, Masters' emotional account highlights the evocative nature of handwriting, going some way to explaining the allure of graphology. In a 2014 lecture, historian Jane Caplan exemplified this persistent appeal by drawing attention to the inelegance of her own signature. Embracing the temptation to associate personality traits with handwriting features, she observed that her signature did not have "the careless flourish of writers who are so important or so busy that they have reduced their signature to a skeletal outline" (Caplan 2014). A connection between handwriting qualities and lifestyle is deeply ingrained in literate culture. For instance, conventional wisdom holds that doctors' handwriting is illegible. In fact, researchers have disagreed about whether this is true. Berwick and Winickoff (1996) state that the writing of doctors is comparable to non-medical executives, whereas Lyons et al. (1998) argue that it is demonstrably worse. The rumours are alarming due to the responsibilities that doctors have: we expect them to take care, and so are surprised by poor-quality writing in prescriptions and medical instructions. Psychologist Rowan Bayne reported to the BBC that graphology's magnetism is partially attributable to the unsophisticated information about personality that *can* be gleaned from handwriting: "[graphology is] very seductive because at a very crude level someone who is neat and well behaved tends to have neat handwriting" (see Duffy and Wilson 2005).

Caplan argues that although writing can be used for identification purposes, it also imparts more subjective ideas of identity. Whilst graphologists cannot claim, convincingly, the objectivity of forensic handwriting analysis, the branches of handwriting analysis "grow from the common tree" of human uniqueness (Caplan 2014). Each discipline engages with the "hand"; a word which "denotes simultaneously the part of our physical body that is most involved in the act of writing, the act of writing itself, and its results on paper" (Caplan 2014). Subjective and objective dimensions of identity are "separable, and yet entangled" in the word "hand" (Caplan 2014). So, I ask, should graphology's lack of credibility guide us to dismiss the more subjective applications of handwriting analysis? The sceptics imply that we should: Nickell distinguishes the work of a questioned-document examiner from that of the graphologist by stating that where the former addresses "a panoply of more or less objective problems", the latter attempts, and fails, to capture the "elusive quality" of a writer's personality (Nickell 1992: 43).

Studies have suggested that graphologists' inferences about these elusive qualities originate from the *content* of handwriting, rather than its form. The subject matter, vocabulary, or other linguistic features of the writing can transmit clues about the writer's personality or state of mind. This idea was proposed by Ben-Shakhar et al. (1986) and Neter and Ben-Shakhar (1989), who argued that the perceived success of graphology might be due to graphologists' observations—conscious or unconscious—about the content of the writing under analysis. We see this in action in *A Life Discarded*, as Masters—unconsciously, or perhaps consciously for dramatic purposes—connects Laura's handwriting features with the personality traits and life events that slowly become clear from the diaries' content.

Psychologists Roy King and Derek Koehler (2000) proposed that this effect is not limited to the content of words, but is also ingrained in the descriptive terminology that graphologists used to describe writing. They investigated a phenomenon known as "illusory correlation" as a contributor to graphology's persistent use to predict personality. In asking participants unfamiliar with graphology to match handwriting with personality profiles, King and Koehler found that "semantic association between words used to describe handwriting features and personality traits was the source of bias in perceived correlation" (King and Koehler 2000: 336).

However, I argue that we do not need to reject entirely the rationale behind graphology—that handwriting features might somehow reflect the more subjective elements of identity—due to the problems inherent in its conventional practice: this risks "throwing the baby out with the bathwater". Thus, I move on to demonstrate how promising research is reconfiguring the investigation into the link between handwriting features and individual psychological difference.

GOODBYE TO "GRAPHOLOGY"? RECONFIGURING THE STUDY OF HANDWRITING FEATURES AND INDIVIDUAL DIFFERENCE

Conventional graphology takes handwriting in isolation, making assertions about the writer's personality based on its features. Its failures arise partially from this approach: any information gleaned from the content of the writing is deemed "cheating". The focus on static features (for instance, the slant of the letters in a single writing sample) eliminate the context of a human life, with its fluctuating psychological, physical, and environmental states. Graphologists have not succeeded in providing significant experimental evidence to demonstrate the link between handwriting features and personality. Academics have not moved to fill this gap; the negative reputation of graphology has discouraged researchers from scientific research in this area. Finally, conventional practice has lacked a comparative element; it has not considered each handwriting sample within the context of a substantial data set.

Digital handwriting recognition is a thriving area of research, and computer scientists have been tempted to apply their methodologies to problems in the under-researched field of graphology. However, digital techniques such as handwriting feature extraction algorithms are tools, and act according to the imperatives of human researchers. Past studies that extract handwriting features and map them to personality traits have struggled to provide convincing evidence for the principles of graphology. They have either employed dubious approaches or provided little to no explanation of their own empirical methodologies (see Górska and Janicki 2012; Gawda 2014; Champa and Kumar 2010, 2011). Despite their failures, these studies illustrate graphology's attractiveness and increasing visibility, even within academic communities. This has been intensified by the proliferation of journals in each subject area, and the increasing availability of open access research online.

More recent, more successful, studies have elected to shed the term "graphology", producing robust, peer-reviewed studies that cautiously investigate writing features that might correspond with the more subjective aspects of identity. For instance, Miguel-Hurtado et al. (2014) combine expertise from psychology and electronics to identify both static and dynamic features in signatures as potential indicators of personality. Explicitly rejecting conventional graphology and adding the further criticism that it ignores the dynamic features of writing (such as writing velocity), they perform a closer scrutiny of how, and which, handwriting features should be measured. In line with previous studies into gender and handwriting, the study finds a connection between gender and writing: "sex classification" and "weight" could be predicted using signature features, especially velocity (see Beech and Mackintosh 2005). However, it also identifies a number of significant correlations between both dynamic and static features and personality features, particularly "conscientiousness" and "perseverance".[2]

Other researchers are shifting the focus away from personality "traits" and towards changeable "states", such as emotions, stress level, and other psychological or psychiatric states. As far back as the 1970s, psychiatrists were interested in handwriting for insight into states of mental health. In 1971, renowned eating disorders specialist Pierre Beumont revealed that he had observed "peculiarities" in the handwriting of patients with anorexia nervosa. Providing illustrative figures, Beumont describes the "extremely small and neat" handwriting of some patients, noting also that in one case a difference could be seen in a sample taken before the onset of the patient's anorexia.

Moving into 2015, Fairhurst, Erbilek, and Li began to quantify the "soft biometrics" of handwriting—information that it contains that is not known to be unique, but is "nevertheless characteristic of an individual"—

with a focus on digital methodologies. Like graphologists, the researchers extract handwriting features to predict "soft" characteristics, but their focus on changeable states, such as levels of happiness, anxiety, or stress levels eradicates the rigidity of conventional graphology. Unlike the prographology articles discussed above, this new research cites studies that suggest that the prediction of emotions from handwriting is possible (Fairhurst et al. 2015; Mutalib et al. 2008). Shifting away from graphology's exclusively human inspection of handwriting, the preliminary study proposes a system for automatically predicting the emotional state of the writer from a handwriting sample.

In a range of pre-determined writing tasks, which were captured using a digitizing tablet with a paper overlay, Fairhurst, Erbilek, and Li found 57% accuracy in a "happy" prediction from the static features of the writing, and a 69.1% accuracy from the dynamic features. When three more features were added to the static feature set, this predictive accuracy increased further. The accuracy levels were less consistent, but not necessarily lower, in unconstrained tasks (where the writing data generated was under the control of the experimental subject). Despite only one sample per person being available in each task, making a longitudinal study impossible, these results are encouraging.

These optimistic studies recommend more in-depth explorations in the area of individual difference, psychological states, and handwriting. Fairhurst, Erbilek, and Li conclude by suggesting further research into the relationship between handwriting features and predictive capability. They propose an investigation into how we can establish ground truth markers for the historical context of emotion, i.e. how do we indicate, reliably, whether a writer was "happy" or otherwise at the time of writing? Finally, they recognize the need to gather more handwriting data to investigate the predictive capability of writing more thoroughly. Importantly, these exploratory studies succeed where the older prographology articles have failed—in avoiding sweeping statements about writing and personality in the absence of convincing evidence about the nature of the link. Instead, the authors display cautious optimism, finally supported by rigorous research. Miguel-Hurtado et al. (2014) suggest that this kind of work could have practical applications in forensics. In the case of a criminal investigation where an offender has left behind handwritten evidence, confident predictions about the person's personality based on his or her writing, could lead to a more targeted search.

CONCLUSION

Unlike phrenology and astrology, which have been consigned to the past as defunct practices, graphology retains popular appeal. Handwriting continues to be seen as a paratextual puzzle waiting to be solved, rich in information about the writer's personality. Moreover, in contrast with the 1990s, when Beyerstein and Beyerstein researched *The Write Stuff*, graphology has extended onto the internet and, to a limited extent, into the methodologies of academic research. Graphology is no longer consigned to "the occult section of local libraries" (Beyerstein and Beyerstein 1992: 15), but instead proliferates on the internet and in the mainstream media, and thus psychologists should continue to investigate its claims. As Nickell explained, "Nonsense rides piggyback on sensible things." Academic research into the validity of the claims of graphologists continues to report against them: there is "no evidence ... to validate the graphological method as a measure of personality" (Dazzi and Pedrabissi 2009).

However, the appeal of graphology—or its foundational idea that individual psychological difference is somehow reflected in handwriting—is itself significant. As Jane Caplan (2014) has shown, the word "hand", as in the phrase "scribal hand", incorporates both objective and subjective concepts of identity. The more that society values individuality, the more alluring becomes the idea that handwriting features are personal, and thus meaningful (Thornton 1998: 140), and, thus, the more we are tempted to investigate the links between handwriting and personality. Furthermore, recent scholarship, distancing itself from conventional graphology, and its sweeping and unsupported statements about personality, is going "back to basics" in the investigation into individual difference and handwriting. With a new focus on rigorous experimental methodologies, significant and representative datasets, and careful peer review, preliminary investigations are optimistic about the link between handwriting features and "soft" biometrics—that is, traits and states that are characteristic of an individual.

NOTES

1. For previous brief discussions of graphology in the context of psychiatric health, see Schiegg and Thorpe (2016), which otherwise focuses on the ways in which handwriting analysis has been used in psychiatric assessment in the early twentieth century (1–24).
2. As measured on the UPPS impulsivity scale (see Whiteside and Lynam 2001.)

BIBLIOGRAPHY

Beech, John R., and Isla C. Mackintosh. 2005. Do Differences in Sex Hormones Affect Handwriting Style? Evidence from Digit Ratio and Sex Role Identity as Determinants of the Sex of Handwriting. *Personality and Individual Differences* 39 (2): 459–468. https://doi.org/10.1016/j.paid.2005.01.024.

Ben-Shakhar, Gershon, Maya Bar-Hillel, Yoram Bilu, Edor Ben-Abba, and Anat Flug. 1986. Can Graphology Predict Occupational Success? Two Empirical Studies and Some Methodological Ruminations. *Journal of Applied Psychology* 71: 645–653. https://doi.org/10.1037/0021-9010.71.4.645.

Berwick, Donald M., and David E. Winickoff. 1996. The truth about doctors' handwriting: A prospective study. *BMJ* 313 (7072): 1657–1658.

Beumont, Pierre. 1971. Small Handwriting in Some Patients with Anorexia Nervosa. *The British Journal of Psychiatry* 119: 349–351. https://doi.org/10.1192/bjp.119.550.349-a. Accessed Jan 2016. Accessed 8 Jan 2015.

Beyerstein, Barry L., and Dale F. Beyerstein. 1992. General Introduction to *The Write Stuff: Evaluations of Graphology, the Study of Handwriting Analysis*, ed. Barry L. Beyerstein and Dale F. Beyerstein, 13–22. Buffalo: Prometheus Books.

Bradley, Nigel. 2005a. Codes of Conduct and Graphology. Paper Presented at the 8th British Symposium of Graphology Proceedings, St. Anne's College, Oxford. http://www.wmin.ac.uk/marketingresearch/graphology/ethicsoxford05eng.htm

———. 2005b. Users of Graphology. *Graphology, the Journal of the British Academy of Graphology* 69: 55–57.

British Institute of Graphologists. 2015. What Is Graphology? http://www.britishgraphology.org/about-british-institute-of-graphologists/what-is-graphology/. Accessed 8 Jan 2015.

Caplan, Jane. 2014. Your Hand: Signatures and Writing. Lecture Presented at Gresham College, London, June 23. http://www.gresham.ac.uk/lectures-and-events/your-hand-signatures-and-handwriting.

Champa, H. N., and K. R. Ananda Kumar. 2010. Automated Human Behavior Prediction Through Handwriting Analysis. Paper Presented at the 2010 First International Conference on Integrated Intelligent Computing (ICIIC), Bangalore, August 5–6, 2010. doi:https://doi.org/10.1109/ICIIC.2010.29.

Champa, H.N., and K.R. Ananda Kumar. 2011. Rule-Based Approach for Personality Prediction Through Handwriting Analysis. *International Journal of Computational Intelligence and Healthcare Informatics* 4: 27–29.

Chartered Institute for Personnel and Development. 1998/2001. Graphology: Quick Facts. London: CIPD.

Crumbaugh, James. 1992. Graphoanalytic Clues. In *The Write Stuff: Evaluations of Graphology, the Study of Handwriting Analysis*, ed. Barry L. Beyerstein and Dale F. Beyerstein, 105–120. Buffalo: Prometheus Books.

Dazzi, Carla, and Luigi Pedrabissi. 2009. Graphology and Personality: An Empirical Study on Validity of Handwriting Analysis. *Psychological Reports* 105: 1255–1268. https://doi.org/10.2466/PR0.105.F.1255-1268.

Douglas, Ian. 2015. Male and Female Brains Are Utter Tosh—Yet We Still Have Men and Women. *The Telegraph*, December 2. http://www.telegraph.co.uk/women/life/male-and-female-brains-are-utter-tosh---yet-we-still-have-men-an/. Accessed 5 June 2017.

Duffy, Jonathan, and Giles Wilson. 2005. Writing Wrongs. *BBC News Magazine*, February 1. http://news.bbc.co.uk/1/hi/magazine/4223445.stm. Accessed 5 June 2017.

Fairhurst, Michael, Meryem Erbilek, and Cheng Li. 2015. Study of Automatic Prediction of Emotion from Handwriting Samples. *IET Biometrics* 4: 90–97. https://doi.org/10.1049/iet-bmt.2014.0097.

Ganz, David. 1999. 'Mind in Character': Ancient and Medieval Ideas About the Status of the Autograph as an Expression of Personality. In *Of the Making of Books: Medieval Manuscripts, Their Scribes and Readers. Essays Presented to M B. Parkes*, ed. P.R. Robinson and Rivkah Zim, 280–299. Aldershot: Scolar.

Gawda, Barbara. 2014. Lack of Evidence for the Assessment of Personality Traits Using Handwriting Analysis. *Polish Psychological Bulletin* 45: 73–79. https://doi.org/10.2478/ppb-2014-0011.

"General Disclaimer." 2015. *Wikipedia, The Free Encyclopedia*. Last Modified December 17, 2015. https://en.wikipedia.org/wiki/Wikipedia:General_disclaimer

Górska, Zuzanna, and Artur Janicki. 2012. Recognition of Extraversion Level Based on Handwriting and Support Vector Machines. *Perceptual & Motor Skills* 114: 857–869. https://doi.org/10.2466/03.09.28.PMS.114.3.857-869.

"Graphology." *Wikipedia, The Free Encyclopedia*. Last Modified May 6, 2017. https://en.wikipedia.org/wiki/Graphology

Greenwood, Chris, and Emine Sinmaz. 2016. Jo Cox's Killer Warned of 'Bloody Struggle' for White Supremacy and Even Signed Off Letters to Pro-apartheid Magazine with 'Racial Regards.' http://www.dailymail.co.uk/news/article-3969826/Jo-Cox-s-killer-warned-bloody-struggle-white-supremacy-signed-letters-pro-apartheid-magazine-racial-regards.html#ixzz4hQGoSmK3. Accessed 5 June 2017.

"How to Read an Article History." 2015. *Wikipedia, The Free Encyclopedia*. Last Modified May 23, 2017. http://en.wikipedia.org/wiki/Wikipedia:How_to_read_an_article_history

Khaleeli, Homa. 2015. Prince Charles Letters: What Does a Graphologist Make of Them? *The Guardian*, March 29. http://www.theguardian.com/uk-news/shortcuts/2015/mar/29/prince-charles-letters-graphologist. Accessed 5 June 2017.

King, Roy N., and Derek J. Koehler. 2000. Illusory Correlations in Graphological Inference. *Journal of Experimental Psychology: Applied* 6: 336–348. https://doi.org/10.1037/1076-898X.6.4.336.

Lockowandt, Oskar. 1992. The Present Status of Research on Handwriting Psychology as a Diagnostic Method. In *The Write Stuff. Evaluations of Graphology. The Study of Handwriting Analysis*, ed. Barry L. Beyerstein and Dale F. Beyerstein, 55–85. Buffalo: Prometheus Books.

Lyons, Ronan, Christopher Payne, Michael McCabe, and Colin Fielder. 1998. Legibility of Doctors' Handwriting: Quantitative Comparative Study. *BMJ* 317: 863–864. https://doi.org/10.1136/bmj.317.7162.863.

Miguel-Hurtado, Oscar, Richard Guest, Sarah V. Stevenage, and Greg J. Neil. 2014. The Relationship Between Handwritten Signature Production and Personality Traits. Paper Presented at the International Joint Conference on Biometrics, Clearwater, September 29–October 2, 2014. https://doi.org/10.1109/BTAS.2014.6996245.

Murray-Browne, Tim. 2016. Foreign Scripts and Familiar Handwriting. *Movement Alphabet*, September 12. http://movementalphabet.com/foreign-scripts-and-familiar-handwriting/. Accessed 5 June 2017.

Mutalib, Sofianita, Roslina Ramli, Shuzlina Abdul Rahman, Marina Yusoff, and Azlinah Mohamed. 2008. Towards Emotional Control Recognition Through Handwriting Using Fuzzy Inference. Paper Presented at the International Symposium on Information Technology, Kuala Lumpur, Malaysia, August 26–28, 2008. https://doi.org/10.1109/ITSIM.2008.4631735.

Neter, Efrat, and Gershon Ben-Shakhar. 1989. The Predictive Validity of Graphological Inferences: A Meta-Analytic Approach. *Personality and Individual Differences* 10: 737–745. https://doi.org/10.1016/0191-8869(89)90120-7.

Newby, Rachel E., Deborah E. Thorpe, Peter A. Kempster, and Jane E. Alty. 2017. A History of Dystonia: Ancient to Modern. *Movement Disorders Clinical Practice* 4: 1–8. https://doi.org/10.1002/mdc3.12493.

Nickell, Joe. 1992. A Brief History of Graphology. In *The Write Stuff: Evaluations of Graphology, the Study of Handwriting Analysis*, ed. Barry L. Beyerstein and Dale F. Beyerstein, 23–29. Buffalo: Prometheus Books.

Rosenblum, Sara, Batya Engel-Yeger, and Yael Fogel. 2013a. Age-Related Changes in Executive Control and Their Relationships with Activity Performance in Handwriting. *Human Movement Science* 32: 363–376. https://doi.org/10.1016/j.humov.2013.08.001.

Rosenblum, Sara, Margalit Samuel, Sharon Zlotnik, Ilana Erikh, and Ilana Schlesinger. 2013b. Handwriting as an Objective Tool for Parkinson's Disease Diagnosis. *Journal of Neurology* 260: 2357–2361. https://doi.org/10.1007/s00415-013-6996-x.

Schiegg, Markus, and Deborah Thorpe. 2016. Historical Analyses of Disordered Handwriting: Perspectives on Early 20th-Century Material from a German Psychiatric Hospital. *Written Communication* 34: 30–53. https://doi.org/10.1177/0741088316681988.

Thompson, Damian. 2015. Charles's 'Spider Letters': The Guardian Falls for the Pseudoscience of Graphology. *The Spectator*, May 13. https://blogs.spectator.co.uk/2015/05/charless-spider-letters-the-guardian-falls-for-the-pseudoscience-of-graphology/. Accessed 5 June 2017.

Thornton, Tamara Plakins. 1998. *Handwriting in America: A Cultural History*. New Haven/London: Yale University Press.

Thorpe, Deborah. 2015. 'I Haue Ben Crised and Besy': Illness and Resilience in the Fifteenth-Century Stonor Letters. *The Mediaeval Journal* 5: 85–108. https://doi.org/10.1484/J.TMJ.5.108526.

"User Contributions. For Geeveraune." 2016. *Wikipedia, The Free Encyclopedia*. https://en.wikipedia.org/wiki/Special:Contributions/Geeveraune. Accessed 5 June 2017.

"User:Tronvillain." 2016. *Wikipedia, The Free Encyclopedia*. Last Modified January 20, 2017. https://en.wikipedia.org/wiki/User:Tronvillain.

Whiteside, Stephen P., and Donald R. Lynam. 2001. The Five Factor Model and Impulsivity: Using a Structural Model of Personality to Understand Impulsivity. *Personality and Individual Differences* 30: 669–689. https://doi.org/10.1016/S0191-8869(00)00064-7.

Williams, Sarah Rhiannon. 2001. *English* Vernacular *Letters* c. 1400–1600: Language, Literacy and Culture. PhD Dissertation, University of York.

CHAPTER 10

Medical Marginalia in the Early Printed Books of University of Glasgow Library

Robert MacLean

The Research Value of Marginalia[1]

Reader marginalia—an increasingly important source of evidence for those studying early books and their readers—is the paratextual feature I will focus on in this final chapter. While the work of Lisa Jardine and Anthony Grafton describing Gabriel Harvey's marginalia (1990) and William H. Sherman's seminal *Used Books* (2008) have been key in demonstrating the possibilities offered by this approach, in the history of science and medicine marginalia has been recognised as a source of evidence for quite some time; see, for example, Vivian Nutton's 1985 article examining annotations by Conrad Gesner and by English physicians Henry Wotton and John Caius in the margins of books they owned. Nutton saw these marginalia as proof of cooperative scholarly networks across Renaissance Europe linking "men of different nationalities and religions in the pursuit of truth". He summed up the value of marginalia evidence rather presciently: "They serve as a salutary reminder that our knowledge, even of the greatest renaissance scientists, can still be enhanced by manuscript discoveries, and that,

R. MacLean (✉)
University of Glasgow Library, Glasgow, UK

© The Author(s) 2018
H. C. Tweed, D. G. Scott (eds.), *Medical Paratexts from Medieval to Modern*, Palgrave Studies in Literature, Science and Medicine, https://doi.org/10.1007/978-3-319-73426-2_10

in this search, one should not neglect *even* the evidence of marginalia" (Nutton 1985: 93–97) [my italicisation].

More than thirty years on from Nutton's article a recent essay by S. Brent Plate—noticing a surge of interest in marginalia studies—identifies two key themes which may go some way towards explaining the increasing scholarly interest. The first concerns the way marginalia attest to active reader engagement with texts—they confirm that readers are not merely passive consumers but informed and opinionated agents, ready to weigh in and give their view. As Plate comments, "The margins are sites of engagement and disagreement: between text and reader and, to stretch it tenuously further, between author and reader ... turning readers into writers, and upsetting the hierarchy of the author as authority" (Plate 2015). And so it is for new media: a whole variety of online forums, news sites and social media encourage "under-the-line" reader comments and reader interaction, thus turning readers into writers. The parallels between new media and early reader marginalia are irresistible and clear to see. This, Plate argues, is one reason why marginalia have become so relevant to current academic discourse.

The other factor Plate identifies—and one particularly relevant to this collection—is the scholarly turn to finally challenge the mind–body dualism which, in many areas of intellectual life, had effaced the human body from study completely. Adrian Johns (1998: 380–443), Katherine Craik (2007) and Helen Smith (2010: 413–432), amongst others, have urged us to reflect on medieval and early modern understanding of the physiology of the eye—theories of extramission and intromission—as a way to attest how acutely conscious early readers were of the corporeality of reading for which, "not only the eyes but also the ears and stomach could be central to the act of textual engagement" (Smith 2010: 415). Reading is a fundamentally embodied act, yet so often we think of it as a simple cognitive process. Encountering marginalia, with their ability to transport us back to a specific time, person and place, demands us to account for readers and their bodies.

While these academic motives are certainly compelling, for me as a librarian, another factor also springs to mind—improving standards of cataloguing and more widespread digitisation which allow researchers to more easily locate annotated copies of works, and if digitised, study them remotely. It is sometimes difficult to determine exactly which came first, the academic interest or the cataloguing; but what cannot be doubted is the importance of good copy-specific cataloguing as an enabler, and in some cases prerequisite for this type of research. When Nutton was writing his article on Gesner in the mid-1980s, researchers relied on the occult world of curator/librarian knowledge and the occasional sheaf binder or annotated printed catalogue to identify potential items of interest. Networked online library catalogues, internet communications protocols, internationally agreed standards on rare book cataloguing, social media, and a willingness on the part of libraries to catalogue books with rich copy-specific metadata have been transformative in opening up previously hidden collections and facilitating the academic study of annotated books. This is a relatively recent phenomenon which is surely also, to some extent at least, fuelling the marginalia *Zeitgeist*.

SURVEY OF ANNOTATED MEDICAL TEXTS IN UNIVERSITY OF GLASGOW LIBRARY

Incunabula[2]

Recent years have seen Special Collections staff at the University of Glasgow Library engaged in projects to enhance early-printed book cata-logue records with rich copy-specific data. Detailed, book-in-hand cata-loguing has brought to light some interesting examples of annotated medical books. The University of Glasgow Library holds 1061 incunab-ula, books printed with movable type during the fifteenth century, one of the largest collections in the UK. The Glasgow Incunabula Project (GIP)[3] has overseen the cataloguing of each of these books with rich copy-spe-cific detail to improve access for researchers. Incunabula are of particular interest to students of marginalia due to the high proportion surviving with early annotation. As Sherman has noted with reference to the Huntington Library holdings, some 60–70% of incunabula are annotated (though, he suggests, the percentage of annotated books extant actually remains high for certain practical genres—including medical texts— throughout the Short Title Catalogue (STC) period (i.e. to 1640)) (Sherman 2008: 9). A staggering 91% of the University of Glasgow's incunabula contain some form of annotation and 64% bear early (i.e. fif-teenth- or sixteenth-century) annotations concerning the text.[4] Over 300 of the University of Glasgow incunabula can be matched to entries in Arnold C. Klebs's bibliography of scientific and medical incunabula (Klebs 1963). Of these, 99 can be considered medical texts,[5] of which some 67% bear early annotations concerning the text.[6]

GIP has grouped annotations into two categories: (1) textual, marginal and interlinear annotations (often responses to the text); and (2) practical annotations, including manuscript pagination, indexing and other additions to help complete a text or help the reader navigate.[7] Other types of mark have also been recorded, including scribbles, pen-trials and reader drawings on margins and endpapers. The following short survey captures some of the variety of annotation found in GIP's medical holdings.

Savonarola, Michele, *Practica medicinae, sive De aegritudinibus.* Venice: Bonetus Locatellus, for Octavianus Scotus, 27 June 1497. [ISTC is00298000]. University of Glasgow Library Sp Coll Bm6-d.12 [GIP S014][8]

This copy of the *Practica medicinae* by Michele Savonarola (1385?–1466?), grandfather of the infamous Dominican friar Girolamo,[9] arrived in Scotland at an early date. It bears the ownership inscription of Mark Jameson (d. 1592), vicar choral of Glasgow, vicar of Kilspindie (Perth and Kinross), and deputy rector at University of Glasgow and was presented to the University by him in 1590. On the rear parchment pastedown are nine lines of medical text in a fifteenth-/sixteenth-century hand containing remedies for "morbus hispanicus"—syphilis. The work contains six full pages of manuscript notes by Jameson (originally bound in as flyleaves but now bound separately in marbled boards) with extracts from the writings of the twelfth-century Salernitan physician, Copho. Sherman has commented that "[m]arginalia can identify other texts a reader associated with or even read alongside a particular book. Cross-references and passages copied verbatim from other books are frequent enough to attest to the widespread practice of what has been called 'extensive' … reading …" (Sherman 2008: 18). These marginalia are a good example of Jameson's extensive reading practice since they follow exactly the text of an early sixteenth-century edition of Copho printed within the medical works of Yūḥannā Ibn Māsawayh (i.e. Mesue) issued from the press of Antoine du Ry for Jacques Giunta in Lyon.[10] Jameson evidently copied verbatim several leaves of Copho from a Giunta Lyon edition of Mesue to augment his Savonarola.[11] Jameson had a clear interest in medicine; not only was he a benefactor of two of the three sixteenth-century hospitals in Glasgow (St. Ninian's, a leper hospital, and St Nicholas's) but—in addition to the Savonarola—he also gifted a number of other medical books to the University in 1590, including two of his annotated herbals and two different sixteenth-century editions of Mesue (neither of which, it should be noted, included the Copho text copied into the Savonarola) (Durkan and Kirk 1977: 173; Durkan and Ross 1961: 119–120).

Those interested in early Scottish owners of medical incunabula should note that Jameson's Savonarola is not the only example of interest in University of Glasgow Library: a *Regimen sanitatis* of Magninus Mediolanensis [GIP M006; ISTC im00052000] is remarkable for having probably been owned by Gilbert Skene (1522?–1599), professor of medicine at King's College, Aberdeen, and physician to James VI of Scotland (a list of recipes for poultices and ointments has been added to an early leaf in a fifteenth-/sixteenth-century hand); while a *Hortus sanitatis* [GIP H41; ISTC ih00488000] survives with Scots language names added by more than one seventeenth-century owner glossing the various woodcuts.[12] Those interested in marginal cross-references to other works as evidence of extensive reading should note that a number of other examples can be found, including a *Regimen sanitatis Salernitanum* dating from ca. 1505 [GIP R003; ISTC ir00081600],[13] the most popular medical work of the incunabular period (De Vos 2013: 672–3). Frequent marginal comments in a sixteenth-century hand cite various authorities including Avicenna and Galen.

Valascus de Tarenta, *Practica, quae alias Philonium dicitur.* Lyon: Johannes Clein, 14 Nov. '1401' [i.e. 1501] [ISTC iv00009000]. University of Glasgow Library Sp Coll Ferguson An-y.22 [GIP V002/1][14]

This medical and anatomical text was written by Valascus de Tarenta (fl. 1382–1418), a French-trained Portuguese physician. Although printed in Lyon, we know that this copy was in the Low Countries at an early date, confirmed by a seventeenth-century ownership inscription of Antwerp's Collegium Medicum on the opening leaf,[15] an early Dutch/Low Countries-style binding, and a manuscript marginal list of medicinal plants in Latin and Dutch. This manuscript list of *materia medica* includes "asarum" (wild garlic) with the corresponding low Dutch translation "mansooren" and "anagallis" (pimpernel), translated into the Dutch "guychelheyl". Both the Latin name, deriving from the Greek "I laugh", and the Dutch, meaning "salvation from madness", hint at its use in treating mental illness and melancholy (Sijs 2010). In comparing early annotations in surviving copies of Hans Caspar Wolf's 1566 *Gynaeciorum libri*, Helen King has noted that vernacular manuscript translations of Latin plant names are an established pattern of marginalia characteristic of a later (seventeenth-century) period of annotation where "the focus shifts to practical use of the texts rather than the scholarly debates within them" (King 2007: 51).

Vernacular manuscript glossing of printed Latin terms can also be found in two different copies of two separate editions of the *Ruralia commoda*, a work on gardening and agriculture by Petrus de Crescentiis (ca. 1233–ca.1320). One copy [GIP C088; ISTC ic00966500] survives with occasional translations into Dutch of the Latin plant names.[16] The other [GIP C087; ISTC ic00966500], which was owned by an unidentified English owner (possibly in Reigate, Surrey) in the mid-sixteenth century, contains frequent marginal annotations in Latin in sixteenth-century hands and, in book six of the volume, many of the printed Latin names for plants have their English equivalent added in a sixteenth-century hand including "lycoryse" (liquorice) beside "De liquiricia", "lyllye" (lily) beside "De lilio" and "letuse" (lettuce) beside "De lactuca".[17] A 1484 *Herbarius latinus* [GIP H006; ISTC ih00062000][18] packed with beautifully hand-coloured woodcuts of different medicinal plants has German synonyms helpfully added by the printer; however, a sixteenth-/seventeenth-century annotator, dissatisfied by the lack of printed index to help navigate the text, has added their own. Such manuscript indexes—created by readers to aid the quick and easy location of different sections on different illnesses and treatments—may well imply that these books were actively being used for practical medicine.

Petrus de Abano, *De venenis*. [Rome: Georgius Sachsel and Bartholomaeus Golsch, ca. 1475] [ISTC ip00440000]. University of Glasgow Library Sp Coll Ferguson An-y.38 [GIP P030][19]

De venenis is a work based on Arabic sources by the Italian philosopher and physician Petrus de Abano (c. 1257–1316) discussing toxins and poisons. This copy's early provenance is unknown but it contains copious Latin annotations in an early hand. On the recto and verso of the opening blank leaf are three columns of Latin terms used for medical conditions, names of medicinal plants and other treatments, all written out neatly. In the margins of the table of contents are annotations in the same early hand, listing medicinal plants and preparations. Many begin with the letters "al" suggesting perhaps the writer had access to a glossary derived from Arabic terms (Fig. 10.1).

These manuscript notes bring together references on toxins and poisons from possibly a number of other sources in one place, for practical use and to avoid loss. This is an example of what Ann M. Blair has called a "treasury" of information, a compilation bringing information and notes together in one place, motivated by that pervasive Renaissance anxiety over loss of knowledge (Blair 2010: 12–13, 61–3, 71–2). As Blair states:

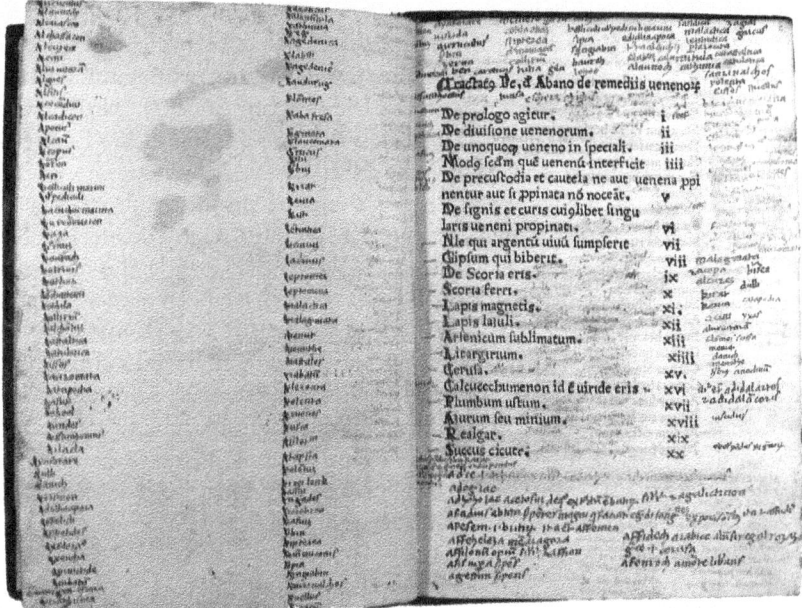

Fig. 10.1 Manuscript list of materia medica with rubricated capital strokes. Petrus de Abano, *De venenis*. [Rome: Georgius Sachsel and Bartholomaeus Golsch, ca. 1475]. Leaves [1v–2r]. University of Glasgow Library Sp Coll Ferguson An-y.38. GIP P030

"Starting in the Renaissance notes were treated less as temporary tools than as long-term ones, worthy of considerable investment of time and effort, of being saved for reuse and in some cases shared with others … Collections of notes were valued as treasuries or storehouses in which to accumulate information even if they did not serve an immediate purpose" (Blair 2010: 63). The creator or early owner's attitude towards these particular manuscript notes—seeing them as long-term and, frankly, anything but marginal or paratextual—is hinted at by the early addition of rubricated capital strokes to both the printed text *and* the manuscript notes to aid navigation and retrieval. The rubrication of the marginalia confers authority and legitimacy on the paratext. It is literally marked as valuable and permanent. But more than that, its whole status as a paratext is arguably transformed: the act of rubrication regularises the look of the book eliding the difference between printed text and handwritten marginalia. Rubrication de-marginalises the marginal and makes text of paratext.

Celsus, Aurelius Cornelius, *De medicina*. Milan: Leonardus Pachel and Uldericus Scinzenzeler, 1481. [ISTC ic00365000]. University of Glasgow Library Sp Coll Hunterian Be.3.27 [GIP C028/1] & Sp Coll Hunterian Bg.3.4 [GIP C028/2][20]

The University of Glasgow holds two copies of this incunable edition of the works of the classical medical authority Aurelius Cornelius Celsus (c. 25 BC–c. 50 BC) whose writing became very popular during the fifteenth century (Pettegree 2011: 297–298). Both copies have been used as practical texts at an early date. The first, once owned by the Dominicans in Mainz, has been foliated by hand and has a rudimentary but unfinished manuscript index added to improve navigation. The second copy bears early marginal annotations in a dull red ink throughout, mostly extracting keywords. In the index beside the relevant sections of text dealing with snake and rabid dog bites, mnemonic drawings have been added of a snake, and a dog (the latter now sadly decapitated due to later cropping by the binder) (Fig. 10.2).

The addition of marginal drawings to augment the text is also found in *Summa conservationis et curationis. Chirurgia*.[GIP S003; ISTC is00033000],[21] a work on hygiene, health and surgery by Guilelmus de

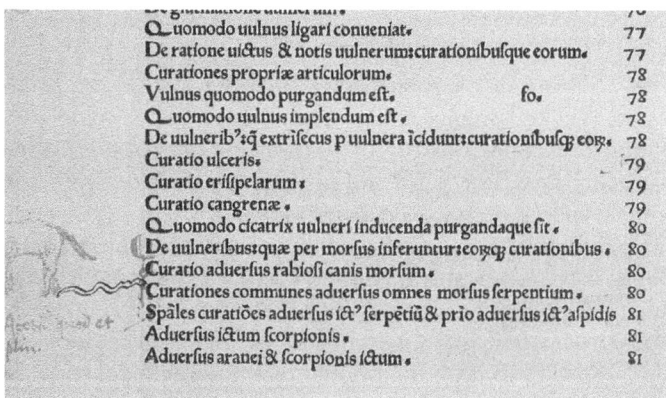

Fig. 10.2 Marginal mnemonic drawing of dog and snake in index. Celsus, Aurelius Cornelius, *De medicina*. Milan: Leonardus Pachel and Uldericus Scinzenzeler, 1481. Leaf [t4v]. University of Glasgow Library Sp Coll Hunterian Bg.3.4. GIP C028/2

Saliceto (ca. 1210–1276 or 1277), an Italian surgeon and cleric. It is the second of two items bound together, the first bearing an ownership inscription of one Phillipus Begardus (fl. 1532). In addition to frequent marginal annotations and an early handwritten contents list, the work includes black pen-and-ink drawings of six surgical instruments, clearly added to illustrate the text. Interestingly the drawings are inserted in spaces apparently deliberately left by the compositor. The absence of printed illustrations of the surgical instruments here seems to imply that woodcut illustrations were unavailable or deemed unsuitable for some reason.

Where woodcut illustrations have been used to illustrate a medical text, marginalia are occasionally associated. Take, for example, the University of Glasgow Library's copy of the lavishly illustrated *Fasciculus medicinae* [GIP K001; ISTC ik00015000][22] by Pseudo Johannes de Ketham, one of the most famous and beautifully illustrated medical works of the fifteenth century. This copy has been captioned in some of the upper margins in a sixteenth-century hand. Manuscript letters of the alphabet (supplementing the letters supplied by the printer) have also been added to the woodcut on a3v as a means of linking the illustrations to the printed text on the facing page.

Glasgow Syphilis Collection[23]

A second recent catalogue enhancement project, this time generously funded by The Wellcome Trust, has seen the University of Glasgow Library catalogue in detail around 250 early printed works on syphilis.[24] A number of these works bear interesting annotations, of which I will provide examples.

Massa, Nicolo, *Liber de morbo Gallico*. [Venice]: [Francis Bindoni, ac Maphei Pasini], [July 1507] [i.e. 1527]. University of Glasgow Library Sp Coll Hunterian Ab.6.31[25]

This copy of Nicolo Massa's work on syphilis—like many of the incunabula featured previously—was owned by the celebrated anatomist, teacher of medicine, physician and collector Dr. William Hunter (1718–1783), whose impressive 10,000-volume library is held by the University of Glasgow. A manuscript note by William Hunter in this copy is particularly interesting, illustrating that his library was not merely for show but was a working research library. A misprint in the colophon of this edition of Massa's text erroneously changing the date 1527 to 1507 had led to it being attributed undue priority by Jean Astruc, the noted eighteenth-century historian of syphilis. Deducing that a mistake had been made, Hunter wrote a long note in his copy, providing evidence for the mistake: "I think there must be an error in the date of this edition: for in the 4th chap. [Massa] tells us of a venereal body which he dissected An[no] 1524; and he hears of guaiacum, which was not known in Italy till 1517" (Maley 2014). Given that Hunter delivered a paper to the Royal College of Surgeons on 14 December 1775 on the origins of Lues Venerea (i.e. syphilis), it is tempting to imagine that his note on the publication date of Massa's *Liber de Morbo Gallico* was written while researching his paper and attempting to establish a timeline of reporting of the disease (Maley 2014).

Astruc, Jean: *De morbis venereis libri sex.* Paris: Guillaume Cavelier [and widow of Paulus-du-Mesnil], [1736]. University of Glasgow Library Sp Coll BG57-b.9[26]

A lengthy and rather gossipy eighteenth- or early nineteenth-century marginal annotation in this work on syphilis recounts the case of a "Noble young lady" who "in a frolic had herself with a many Breeches who had been afflicted with lues [i.e. syphilis]", fell ill with "a most afflicting ulceration of the pudendum" and "her modesty preventing her from divulging", died. This work bears a bookplate suggesting it was in the University of Glasgow Medical Class Library at an early date, so the marginalia may have been made by a student. The subject matter and the tone are an important reminder that attending to the evidence of reader marginalia alone in addressing reading history might result in a gendered approach. As Heidi Brayman Hackel has noted, early modern women readers tended not to annotate; "reading-only" literacy and cultural expectations of feminine modesty militated against it. Therefore, a focus on marginalia potentially elides women readers from a full and accurate history of reading (Brayman Hackel 2005: 196–255). Brayman Hackel (2005: 254) suggests that "[m]any gentlewomen displayed the importance of their book ownership in [other] material ways: in elaborate bindings, careful catalogues, commissioned portraits, gift exchanges and final bequests".

On the question of gender, an intriguing annotation can be found in another medical text recently catalogued in detail: a 1631 second edition of Helkiah Crooke's *Mikrokosmographia* [ESTC S107279].[27] A woodcut diagram of a female torso showing "the breast of a woman, with the skin flayed off", and with internal and reproductive organs exposed, has been amended by hand. The annotator has added a woman's head to the dismembered torso, with smiling face, and a bow tying her hair in place. Is this a reader's attempt to add some humanity to the objectified torso?

Who was the annotator? We just do not know, though the addition seems likely to have taken place in the eighteenth century or earlier. For a more detailed consideration of this example, see Harry Newman's chapter "[P]Rophane Fidlers': Medical Paratexts and Indecent Readers in Early Modern England" in this volume.

Hunter, John, *A treatise on the venereal disease*. London: Sold at No. 13, Castle-Street, Leicester-Square; and by Mr. G. Nicol, Pall-Mall; and Mr. J. Johnson, St. Paul's Church-Yard, [1788]. University of Glasgow Library Sp Coll 83.b.15[28]

An amusing sequence of late eighteenth- or early nineteenth-century student notes can be found in this copy of John Hunter's work on venereal disease, which was in the University Library at an early date. Objecting to a medical case study in the text which commented that a 16-year-old boy had "fine healthy children", a student responded in the margins, "The Devil! how many could he have? he was but sixteen years of age!" Beneath the annotation, a second reader has responded, correcting the first, culminating with a witty put-down, "You are an Ass—a pretty fellow to be a student of medicine. Does not your Jackass brains perceive that on the previous page this patient was seventeen years of age while under treatment ... and that many years must have elapsed ... till this case...? O thou donkey! Thou beast of burthen!—Has your Mother any more hopeful sons?" This entertaining marginal exchange, with reader–author and reader–reader dialogue, is—in form and tone—similar to one of the "below the line" comment threads so common to new media (Fig. 10.3).

THE IMPORTANCE OF CATALOGUING PROJECTS

Descriptive metadata-rich copy-specific cataloguing of rare books is a slow and resource-intensive process. It requires skilled cataloguers with expert knowledge and well-developed language and palaeography abilities. It is expensive and difficult to fund. A recent Research Libraries UK survey indicated that rare book cataloguing remains a huge problem for UK special collections libraries with 13 million volumes—nearly 1 in 5—rare books remaining completely uncatalogued (RLUK 2012, 6). And this figure does not even factor in that a large proportion of the 80% which *have* been catalogued lack provenance histories, binding, annotation and other copy-specific descriptions. In an age of austerity, and with Higher Education Library budgets shouldering the burgeoning cost of key research journal subscription, funding rare book cataloguing is commonly

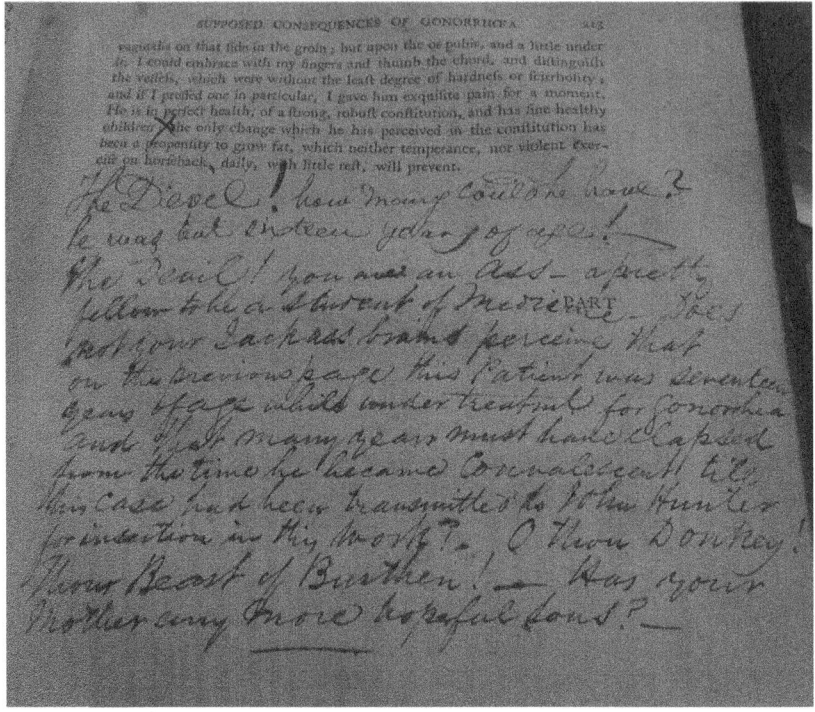

Fig. 10.3 Late eighteenth- or early nineteenth-century student dialogue in the margins. John Hunter, *A treatise on venereal disease*. London: Sold...by Mr. G. Nicol..., 1788. Page 213. University of Glasgow Library Sp Coll 83.b.15

not prioritised. Yet few other funding sources exist, since external funding bodies often do not look favourably on cataloguing project bids, seeing this activity as something core which should be funded directly from libraries' budgets.

The Glasgow Incunabula Project (GIP) has taken seven years and thousands of staff hours to complete. Too expensive to fund entirely from already stretched University Library resources, and having failed to attract external funding, the project has taken place largely through the generosity of the primary investigator, Jack Baldwin. We were very fortunate that The Wellcome Trust saw the value in our collection of early syphilis books and generously funded a six-month cataloguing project to improve researcher access.

Worldwide, only a fraction of early printed books have been catalogued with the level of rich copy-specific detail which permits viable research into marginalia. While some important projects are ongoing to address this lack,[29] until many more early printed books are examined and described, there are certain types of research question which will be very difficult to address. Take, for example, William H. Sherman's assertion—based on books held by the Huntington Library—that book annotation remains common for certain genres of practical books, including medical texts, up to the middle of the seventeenth century (Sherman 2008: 9); this may well be true for the Huntington collections, but can we say this more broadly? Or Helen King's claim based on examining copies of the *Gynaeciorum libri,* that vernacular manuscript translations of Latin plant names are characteristic of a later (seventeenth-century) annotation (King 2007: 51). While this might be true for surviving copies of the *Gynaeciorum libri*, is it true for all such annotations in early medical works? Or even Heidi Brayman Hackel's belief that early modern women readers did not annotate their books but displayed their interest in other material ways (Brayman Hackel 2005: 254)—can we safely say this when so many early printed books remain poorly catalogued and hidden to researchers?

It is important that marginalia researchers recognise the difficult funding landscape that libraries face in trying to enhance rare book catalogue records. Funding avenues closed to librarians are often open to university-based researchers and funders may be receptive to an argument made by a researcher that detailed cataloguing is a necessary prerequisite to answering their research question. Therefore, one possible path forward is to build rare book catalogue-enhancement funding into wider researcher-led bids at the outset. If you are hoping to embark on a research project focusing on marginalia, it would be wise to contact the appropriate special collections librarians at an early stage in planning to discuss the merits of collaboration.

Reader marginalia are one of the most valuable forms of surviving evidence of early reading. They offer great potential to researchers attempting to understand how early modern people read and engaged with books as physical objects. And with so many early published works focusing on medicine and health, marginalia offer particular opportunities for those interested in medical history. Detailed cataloguing projects like GIP are creating finding aids which make it easier for researchers to locate and access early annotated books. And yet large numbers of books remain entirely hidden from view to researchers, either completely uncatalogued or catalogued without copy-specific metadata.

Notes

1. I will use the term marginalia and annotation interchangeably throughout but for a discussion on terminology, see William H. Sherman. 2008. *Used Books: Marking Readers in Renaissance England* (Philadelphia: University of Pennsylvania Press), 20–24.
2. The books featured here were first identified and described by Jack Baldwin, primary investigator on Glasgow Incunabula Project, to whom I owe many thanks. And many thanks to Julie Gardham, Senior Librarian and Head of Special Collections, University of Glasgow Library, for her invaluable input.
3. University of Glasgow Library. 2017. "Glasgow Incunabula Project". Accessed June 23 2017, http://www.gla.ac.uk/services/incunabula/
4. The percentage would be even higher if we consider practical annotation, e.g. manuscript foliation, manuscript signatures, manuscript indexes, etc.
5. GIP added Library of Congress Subject headings (LCSH) beginning "Medicine ..." to 89 editions [i.e. 99 copies]. However, it should be noted that multiple texts—occasionally on differing topics—can be included in a single incunabulum; LCSH are not necessarily added to reflect every single text in a work, so this count may underrepresent the true number of "medical" incunables held. Also, this figure does not include non-medical works which include "medical" marginalia. See, for example, a 1489 Strasburg *Biblia Latina* [GIP B062; ISTC ib00588000] owned in the sixteenth century by a Benedictine monastery in Ettenheimmünster, Baden-Württemberg, which includes a five-line inscription in German at head of the initial leaf in a sixteenth-century hand containing a medical recipe to treat scalds. See University of Glasgow Library [i.e. UofG]. 2017. Sp Coll Euing Dt-d.14: "Biblia Latina", accessed June 23, http://www.gla.ac.uk/services/incunabula/a-zofauthorsa-j/dt-d.14/
6. Once again, the percentage would be even higher if we consider practical annotation.
7. For a general discussion on taxonomies of marginalia, see Sherman, William H. 2008. *Used Books: Marking Readers in Renaissance England.* (Philadelphia: University of Pennsylvania Press), 16–17. For another suggested typology of marginalia in incunabula, see University of Oxford. 2017. "Reading Practices", *University of Oxford 15cBooktrade*, accessed June 23 http://15cbooktrade.ox.ac.uk/reading-practices/
8. UofG. 2017. Sp Coll Bm6-d.12 "Savonarola, Michael: Practica medicinae, sive De aegritudinibus", accessed June 23, http://www.gla.ac.uk/services/incunabula/a-zofauthorsa-j/bm6-d.12/
9. Girolamo Savonarola (1452–1498) was a preacher and prophet who came to prominence in late fifteenth-century Florence for his attacks on corruption within the Church and his calls for religious reform.

10. Yūḥannā Ibn Māsawayh (i.e. John Mesue) [Pseudo-]. 1523. *D[omi]ni Mesue vita;*
    Doctoru[m] artis peonie cognomina. Canones vniuersales diui Mesue de consolatione medicinarum ... Lyon: Antoine du Ry for Jacques Giunta, cf. a1r-a5r (see: Baudrier VI, 112). For a discussion of Yūḥannā Ibn Māsawayh, the problems on establishing authorship, and the medieval and early modern popularity of these works, see Paula De Vos (2013: 667–712).

11. It is worth noting that Giunta issued further editions of this work in subsequent years with much the same typesetting (see e.g. Baudrier VI, 142 published in 1531); it is therefore uncertain from exactly which edition Jameson copied this text.

12. UofG. 2017. Sp Coll Hunterian By.3.31 "Magninus Mediolanensis: Regimen sanitatis", accessed June 23, http://www.gla.ac.uk/services/ incunabula/a-zofauthorsa-j/by.3.31/ and University of Glasgow Library. 2017. Sp Coll Bm4-e.2 "Hortus sanitatis.", accessed June 23, http:// www.gla.ac.uk/services/incunabula/a-zofauthorsa-j/bm4-e.2/

13. UofG. 2017. Sp Coll Hunterian Bw.3.3 "Regimen sanitatis Salernitanum", accessed June 26, http://www.gla.ac.uk/services/ incunabula/a-zofauthorsa-j/bw.3.3/

14. UofG. 2017. Sp Coll Ferguson An-y.22 "Valascus de Tarenta: Practica, quae alias Philonium dicitur", accessed June 26, http://www.gla.ac.uk/ services/incunabula/a-zofauthorsa-j/An-y.22&Ck.2.1/

15. The Collegium Medicum was founded by the City of Antwerp in 1659 incorporating the medical library of Antwerp resident Dr. Jean Ferreulx (1557–1627). This book is listed in the original donation from Ferreulx to the city. Many thanks to Steven Van Impe of the Hendrik Conscience Library, Antwerp, for this information.

16. UofG. 2017. Sp Coll Hunterian Bh.3.25 "Crescentiis, Petrus de: Ruralia commoda", accessed June 26, http://www.gla.ac.uk/services/ incunabula/a-zofauthorsa-j/bh.3.25/

17. UofG. 2017. Sp Coll Hunterian Bw.2.17 "Crescentiis, Petrus de: Ruralia commoda", accessed June 26, http://www.gla.ac.uk/services/ incunabula/a-zofauthorsa-j/bw.2.17/

18. UofG. 2017. Sp Coll Hunterian Bw.3.5 "Herbarius latinus", accessed June 26, http://www.gla.ac.uk/services/incunabula/a-zofauthorsa-j/bw.3.5/

19. UofG. 2017. Sp Coll Ferguson An-y.38 "Petrus de Abano: De venenis", accessed June 26, http://www.gla.ac.uk/services/incunabula/ a-zofauthorsa-j/an-y.38/

20. UofG. 2017. Sp Coll Hunterian Be.3.27 & Sp Coll Hunterian Bg.3.4 "Celsus, Aurelius Cornelius: De medicina", accessed June 26, http:// www.gla.ac.uk/services/incunabula/a-zofauthorsa-j/be.3.27%20&%20 bg.3.4/

21. UofG. 2017. Sp Coll Hunterian Bh.1.17 (item 2) "Saliceto, Guilelmus de: Summa conservationis et curationis. Chirurgia", accessed June 26, http://www.gla.ac.uk/services/incunabula/a-zofauthorsa-j/bh.1.17b/
22. UofG. 2017. Sp Coll Hunterian Ds.2.2 "Ketham, Johannes de [pseudo-]: Fasciculus medicinae", accessed June 26, http://www.gla.ac.uk/services/incunabula/a-zofauthorsa-j/ds.2.2/
23. The books featured here were first identified and described by Sonny Maley, Wellcome Syphilis Project officer, to whom I owe many thanks.
24. University of Glasgow Library. 2017. "Syphilis Collection", accessed June 26, http://www.gla.ac.uk/services/specialcollections/collectionsa-z/syphiliscollection/
25. UofG. 2017. Sp Coll Hunterian Ab.6.31 "Nicolai Massa Veneti:: artiu[m] & medicine doctoris: Liber de morbo Gallico:.", accessed June 26, http://eleanor.lib.gla.ac.uk/record=b3082690
26. UofG. 2017. Sp Coll BG57-b.9 "De morbis venereis libri sex", accessed June 26, http://eleanor.lib.gla.ac.uk/record=b3075134
27. UofG. 2017. Sp Coll Hunterian Aa.2.19 "Mikrokosmographia", accessed June 26, http://eleanor.lib.gla.ac.uk/record=b1606622
28. UofG. 2017. Sp Coll 83.b.15, "A treatise on the venereal disease", accessed June 26, http://eleanor.lib.gla.ac.uk/record=b3092972
29. See, for example, the Consortium of European Research Libraries (CERL) *Material Evidence in Incunabula* database; the collaborative *Annotated Books Online* (ABO) project; and the collaborative *Archaeology of Reading in Early Modern Europe* project.

BIBLIOGRAPHY

Annotated Books Online. 2017. *A Digital Archive of Early Modern Annotated Books.* http://www.annotatedbooksonline.com/. Accessed 27 June.

Archaeology of Reading. 2016. *The Archaeology of Reading in Early Modern Europe.* Last modified, September. http://archaeologyofreading.org/. Accessed 27 June 2017.

Baudrier, Henri Louis. 1895–1921. *Bibliographie lyonnaise : recherches sur les imprimeurs, libraires, relieurs et fondeurs de lettres de Lyon au XVIe siècle.* Lyon: Librairie Ancienne d'Auguste Brun.

Blair, Ann M. 2010. *Too Much To Know: Managing Scholarly Information Before the Modern Age.* New Haven/London: Yale University Press.

Brayman Hackel, Heidi. 2005. *Reading Material in Early Modern England: Print, Gender, and Literacy.* Cambridge: Cambridge University Press.

CERL. 2016. *Material Evidence in Incunabula.* Last modified, June 30, 2016. https://www.cerl.org/resources/mei/main

Craik, Katherine. 2007. *Reading Sensations in Early Modern England.* Basingstoke: Palgrave Macmillan.

De Vos, Paula. 2013. "The Prince of Medicine" Yūḥannā Ibn Māsawayh and the Foundations of Western Pharmaceutical Tradition. *ISIS: A Journal of the History of Science Society* 104: 667–712.

Durkan, John, and James Kirk. 1977. *The University of Glasgow 1451–1577.* Glasgow: University of Glasgow Press.

Durkan, John, and Anthony Ross. 1961. *Early Scottish Libraries.* Glasgow: John S. Burn and Sons.

Jardine, Lisa, and Anthony Grafton. 1990. 'Studied for Action': How Gabriel Harvey Read His Livy. *Past and Present* 129: 30–78.

Johns, Adrian. 1998. The Physiology of Reading. In *The Nature of the Book: Print and Knowledge in the Making*, 380–443. Chicago/London: University of Chicago Press.

King, Helen. 2007. *Midwifery, Obstetrics and the Rise of Gynaecology: The Use of a Sixteenth-Century Compendium.* Aldershot: Ashgate.

Klebs, Arnold C. 1963. *Incunabula Scientifica et medica.* Hildesheim: George Olms.

Maley, Sonny. 2014. Unexpected Notes on Syphilis, *University of Glasgow Library Blog.* December 5. https://universityofglasgowlibrary.wordpress.com/2014/12/05/unexpected-notes-on-syphilis/. Accessed 27 June 2017.

Nutton, Vivian. 1985. Illustrations from the Wellcome Institute Library: Conrad Gesner and the English Naturalists. *Medical History* 29: 93–97.

Pettegree, Andrew. 2011. *The Book in the Renaissance.* New Haven/London: Yale University Press.

Plate, S. Brent. 2015. Marginalia and Its Disruptions. *Los Angeles Review of Books.* 16 December. https://lareviewofbooks.org/article/marginalia-and-its-disruptions. Accessed 4 July 2016.

RLUK. 2010. *Hidden Collections: Report of the Findings of the RLUK Retrospective Cataloguing Survey in Association with The London Library.* http://www.rluk.ac.uk/wp-content/uploads/2014/02/RLUK-Hidden-Collections.pdf. Accessed 27 June 2017.

Sherman, William H. 2008. *Used Books: Marking Readers in Renaissance England.* Philadelphia: University of Pennsylvania Press.

Smith, Helen. 2010. "More swete vnto the eare/than holsome for ye mynde": Embodying Early Modern Women's Reading. *Huntington Library Quarterly* 73 (3): 413–432.

University of Glasgow Library. 2017a. *Glasgow Incunabula Project.* http://www.gla.ac.uk/services/incunabula/. Accessed 23 June.

———. 2017b. *Syphilis Collection.* http://www.gla.ac.uk/services/specialcollections/collectionsa-z/syphiliscollection/. Accessed 26 June 2017.

University of Oxford. 2017. Reading Practices, *University of Oxford 15cBooktrade.* http://15cbooktrade.ox.ac.uk/reading-practices/. Accessed 23 June.

van der Sijs, Nicoline, ed. 2010. Guichelheil (Anagallis arvensis), *Etymologiebank.nl.* http://www.etymologiebank.nl/trefwoord/guichelheil. Accessed 26 June 2017. Reference derives from H. Kleijn. 1970. *Planten en hun naam.* Amsterdam: Meulenhoff.

Correction to: "Nonsense Rides Piggyback on Sensible Things": The Past, Present, and Future of Graphology

Deborah Ellen Thorpe

CORRECTION TO:

Chapter 9 in: H. C. Tweed, D. G. Scott (eds.), *Medical Paratexts from Medieval to Modern*, Palgrave Studies in Literature, Science and Medicine, https://doi.org/10.1007/978-3-319-73426-2_9

The chapter ""**Nonsense Rides Piggyback on Sensible Things": The Past, Present, and Future of Graphology**" has been made Open Access under a CC BY 4.0 license.

The updated online version of this chapter can be found at
https://doi.org/10.1007/978-3-319-73426-2_9

© The Author(s) 2018   E1
H. C. Tweed, D. G. Scott (eds.), *Medical Paratexts from Medieval to Modern*, Palgrave Studies in Literature, Science and Medicine,
https://doi.org/10.1007/978-3-319-73426-2_11

INDEX[1]

---

[1] Note: Page numbers followed by 'n' refer to notes.

© The Author(s) 2018   175
H. C. Tweed, D. G. Scott (eds.), *Medical Paratexts from Medieval to Modern*, Palgrave Studies in Literature, Science and Medicine, https://doi.org/10.1007/978-3-319-73426-2

The manufacturer's authorised representative in the EU is Springer
Nature Customer Service Centre GmbH, Europaplatz 3, 69115 Heidelberg,
Germany. If you have any concerns regarding our products, please
contact ProductSafety@springernature.com

Printed and bound by CPI Group (UK) Ltd, Croydon, CR0 4YY
27/04/2026
02097631-0001